THE HISTORY OF TIME
UNIVERSE

BY DAVID BLAIR & GEOFF CODY

Title: The History of Time: UNIVERSE

Authors: David Blair & Geoffrey Cody

ISBN: 978-0-646-58069-2

Format: Paperback

Publication Date: 07/2012

© 2012, David Blair and Geoffrey Cody

Published by: Gravity Discovery Centre

Graphic Design: www.lushartdesign.com.au / karen@lushartdesign.com.au

Printed by: www.createspace.com

Contents

Timeline to the Present

Discoverers and Discoveries

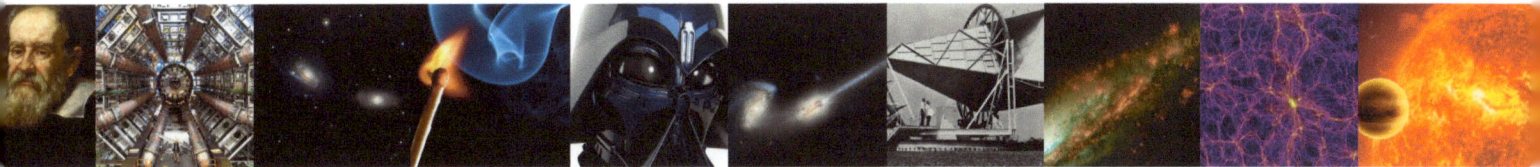

The Future of the Universe

About the Authors

Introduction

All of us share a common yearning to understand our place in the universe. For countless millennia we looked at the stars, the land and ourselves, and wove explanations that involved unseen forces and unseen beings. Every explanation was a rich and detailed tapestry, a story that linked our existence to cosmic powers beyond our understanding. The forces and beings were all endowed with strong human emotions, from desire to anger, and were believed to have powerful influences over everything including the individual inhabitants of their world.

Over the past few centuries a new story of the universe has emerged. It has been gradually uncovered through the meticulous efforts of thousands of people. It is the outcome of a new way of looking at the world: one that combines precision measurement with logic and scepticism. Scepticism is an approach to knowledge that acknowledges human fallibility. It takes the concept of truth off its pedestal of absolute knowledge (usually revealed by a sage or a prophet), and turns it into an approximation that can be argued and debated.

The new story of the universe has slowly emerged out of conceptual struggles, acrimonious debates and mental wars. Measurements with better and better instruments, from telescopes to microscopes, have provided ever finer data with which to test theories, while scepticism has allowed the theories of even the most venerable old teachers to be discarded. The revelations in the new story of the universe create a new

tapestry much richer than the old explanations, and one that encompasses mind-numbing distances in space and time. What is revealed is a violent and impersonal universe, but one which appears to be perfectly designed to allow it to create observers – ourselves – with the astonishing ability to look at the universe as a whole and understand it.

The new story of the universe is unfinished. The quantum processes that structure matter are weird, leaving questions about the basic nature of reality. We do not know why, nor exactly how, the processes that lead from inanimate atoms to self aware organisms originate. We do not know why the universe began and whether it will repeat itself. We suspect that the universe might be teeming with life, but we are entirely ignorant of life beyond the solar system. We find that more than 90% of the stuff of the universe is unknown, partly an unknown type of matter called dark matter, and partly an unknown force acting on space, called dark energy.

While we have an idea of the vastness of the visible universe, we are completely uncertain of its multiplicity. Is our visible universe but a dust grain in an incomprehensibly larger cosmos?

In the few years since the *Timeline of the Universe* was opened we were constantly asked "Who wrote those poems?" "Are they available in a book?". Eventually we were inspired to create this book. Here you will find an image and brief explanatory text to describe every stage in the universe, and a poem that tries to encapsulate the ideas. While each poem is attributed to one of us, we have actually collaborated and argued and used the best of scientific scepticism to help each of us improve our poems. This book is a record of our warm collaboration.

David Blair and Geoff Cody
January 2012

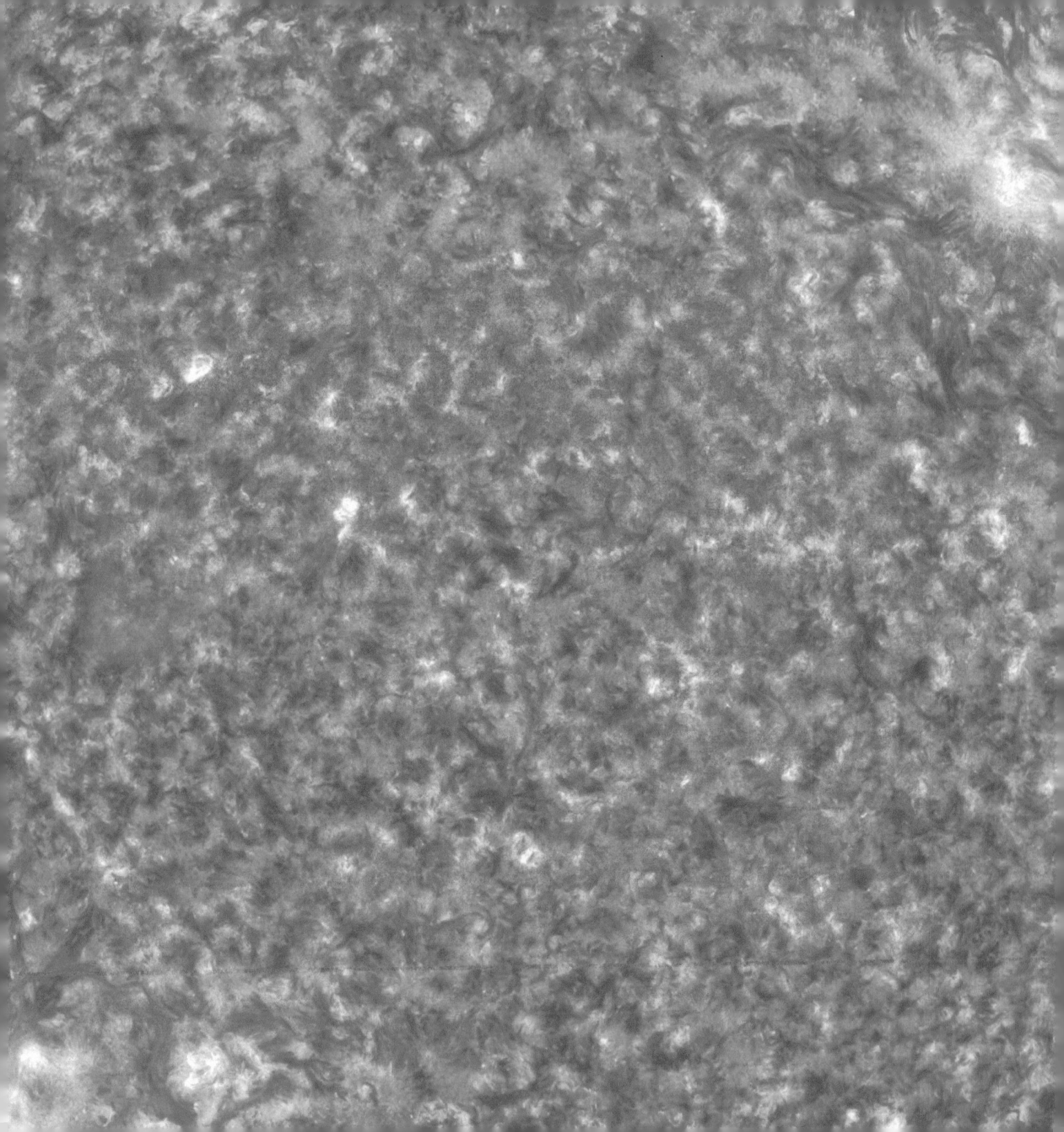

The Mystery of the Big Bang

Time: 10^{-35} seconds after the big bang, the moment of creation, 13.7 billion years ago.

The universe was almost infinitely hot, and almost infinitely dense. It was not an explosion in empty space but an explosion of space. Why and how it occurred is a mystery. An immensely powerful force – some kind of *dark energy* – inflated the universe like a rapidly inflating balloon. After a momentary burst of inflation, it began coasting like the sparks of a giant firework.

Unlike fireworks, it is important to understand that the big bang was not something that happened at one place in space. The ants' space is the rubber surface. Their two dimensional space is expanding away from them in all directions as the balloon expands. Our experience of the big bang is like the ants' experience of the expanding balloon.

The entire universe was hot and compressed. An explosively expanding balloon is the best analogy for the big bang. Using this analogy we must think of ourselves as flat two dimensional ants looking out on the rubber surface of the balloon. There is no centre, and no particular place it expands from.

The seething hot surface of the Sun captures the idea of the big bang as experienced by an ant in which its universe is the sun's two dimensional surface.

Emergence

In the beginning
there was no beginning

from no law
law emerged

from no-time and no-space
time and space emerged

from chaos
order emerged

from radiation
matter emerged

from non-existence
existence emerged

Forces Emerge

Time: before the first microsecond, 13.7 billion years ago.

The universe is structured by forces defined by laws of physics that appear to be obeyed universally across time and space. On the smallest scale two types of nuclear forces make atomic nuclei. On a bigger scale the electromagnetic force makes atoms, molecules and life. On the biggest scale gravity creates planets, stars and galaxies.

In the beginning, for the tiniest fraction of a microsecond, in the indescribable heat of the big bang, a single repulsive super force drove the expansion of space. As the universe expanded, attractive forces emerged. First gravity emerged from the other forces, followed by the strong nuclear force. Lastly the electromagnetic force separated from the weak nuclear force. This last separation has been observed in the laboratory but all the rest is conjecture.

Creation was an act of symmetry breaking, the emerging forces broke the symmetry of the universe. In the first microsecond the universe was a thick soup of matter, antimatter and radiation. The symmetry between matter and antimatter was broken by just one part in a billion, allowing a tiny remnant of matter to be left over when the vast majority of it annihilated to create radiation.

Recently a force called dark energy was found to be powering an accelerating expansion of the universe. Dark energy may be a very weak remnant the force that drove the big bang. It appears to act on the volume of space. The ultimate fate of the universe depends on its precise but still unknown properties.

The Penrose tiling at the Cosmology Gallery of the Gravity Discovery Centre in Australia is an example of broken symmetry. Penrose tilings can be created with perfect symmetry or with broken fivefold symmetry as we see here.

Law enforces

From unification
four forces emerge
rulers of the universe

hidden in perfect laws

matter
dark matter
and radiation

obedient servants
symmetry breakers

The First Atomic Nuclei Form

Time: before the first three minutes, 13.7 billion years ago.

In the minutes after the universe began there was a brief moment when protons and neutrons could combine to make the nuclei of helium, deuterium and lithium. Earlier it was too hot and the nuclei were all destroyed. A few minutes later it was too cool and there was not enough energy to stick them together.

The amazing agreement between the measured and predicted amounts of hydrogen, helium, and deuterium in the universe convinced physicists that there really was a big bang 13.7 billion years ago.

At this time the mass of the universe was dominated by the mass energy of electromagnetic radiation called gamma rays. The atoms were a tiny fraction.

Another form of matter was also present – dark matter. This matter did not interact with the atoms. It was quite invisible, but its inertia would play a key role in determining the structure of the universe in the future. Dark matter is still a mystery. It is only known to exist because of its gravitational effects.

Hydrogen balloons floating free.

Broken symmetry

In fiery equilibrium
matter and radiation coexist
cooling fast
antimatter annihilates with matter
a tiny condensation of matter
remains
hydrogen and helium

Cooling Expansion

Time: about 10,000 years after the big bang, 13.7 billion years ago.

Expansion causes cooling. As the universe expanded, the radiation cooled rapidly, giving up its energy to gravity. Matter and dark matter carried on for the ride, swamped by a sea of radiation.

At 10,000 years of age, the still violently expanding universe was a blue hot fog. The almost uniform gas had a temperature of more than 100,000 degrees. At this temperature electrons cannot be retained by atomic nuclei so neutral atoms cannot exist. The electrons roam free amongst a cloud of atomic nuclei. In this state called plasma, light is constantly absorbed and re-radiated by the free electrons, and like the sun, it is opaque. Most of the mass and the energy in the universe was in the form of photons like X-rays and ultraviolet light.

Blue hot fog

Blue hot fog
powered by heat
expanding
fighting gravity
cooling
slowing
losing the fight

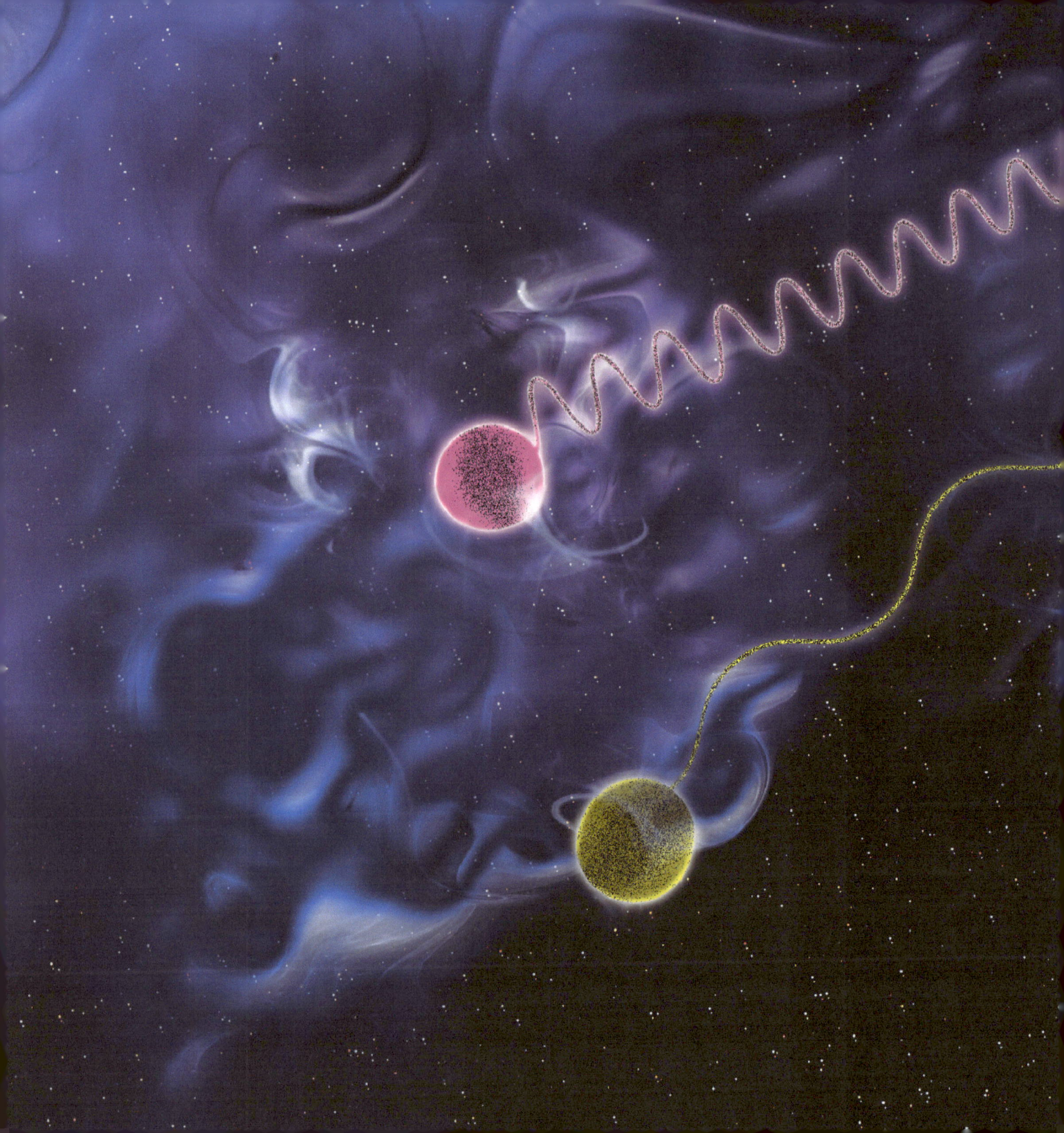

Photons

Einstein proved that light and all electromagnetic waves come in small bundles of electromagnetic energy called photons. All photons have energy and travel at the speed of light, about 300,000 kilometres per second. We see because photons trigger sensory cells in our eyes that send nerve pulses to our brain. Our mobile phones emit and receive streams of invisible radio photons.

The vast majority of the photons in the universe are microwave photons. They were once part of the blue hot fog in the early Universe. Today after 13.7 billion years of expansion they have lost most of their energy to gravity. There are about 10 billion of these microwave photons for every particle of matter in the universe.

Artist's impression of photons in space.

Betrayal

Photons
particles of light
ten billion for every atom
once powered the universe
today a ghostly remnant
secrets from the past
betrayed

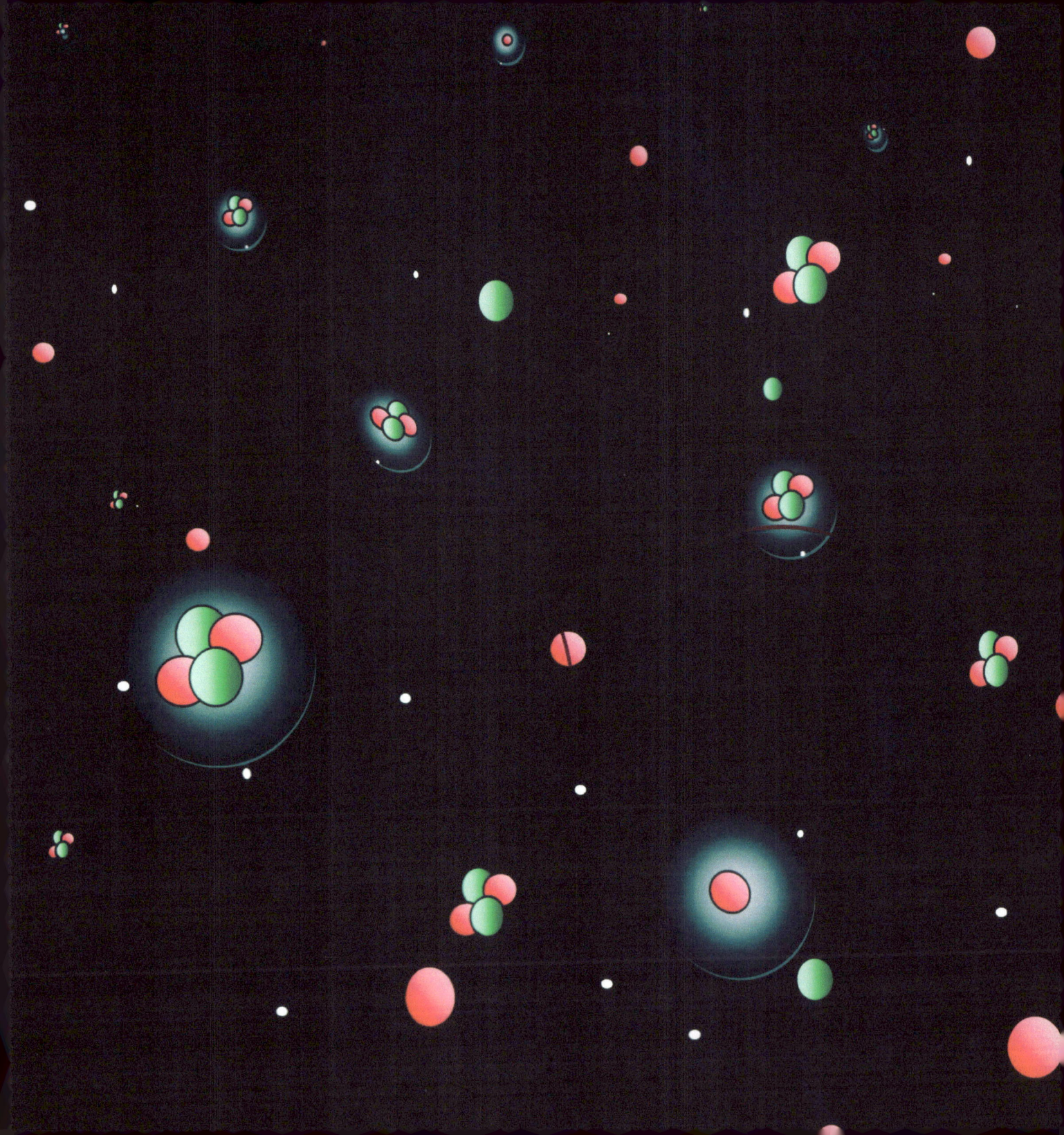

The First Neutral Atoms Form

Time: about 380,000 years after the big bang, 13.7 billion years ago.

The expansion of the hot fog cooled it to a few thousand degrees. This was cool enough that electrons could be captured into stable orbits around the hydrogen and helium nuclei. This changed the gas from being an opaque plasma like the sun, to transparent gas like the air. Suddenly light could move freely through the gases. Suddenly the fog was lifted! Space continued to cool from yellow to orange like the inside of a hot furnace. Then to the black we see today.

By this time the mass of the radiation was greatly diminished. The mass of the universe was now dominated by matter in two forms: the hot gas of hydrogen and helium, plus the dark matter, which contributed about 7 times the mass of the gas. This dark matter would go on to determine how the universe evolved.

Artist impression of free electrons neutralising the hydrogen and helium nuclei.

Sky

Yellow hot fog
suddenly clears
transparent yellow sky
cooling to red
to black
as space expands

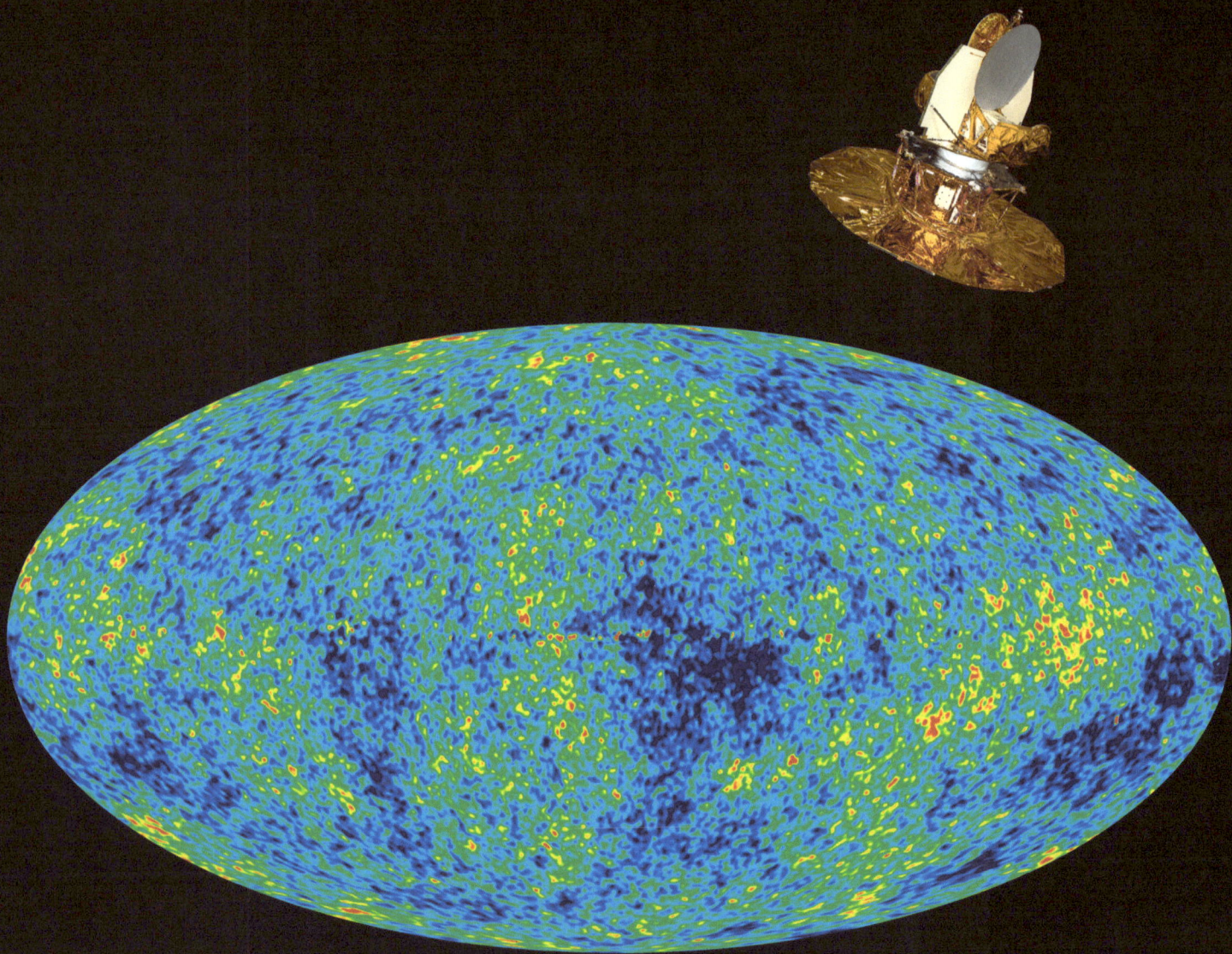

The Surface of
Last Scattering

Time: about 380,000 years after the big bang, 13.7 billion years ago.

After neutral atoms had formed the photons travelled freely through space like light through the air. Today, in every direction, we see those photons. They have been stretched a thousand-fold to become microwaves as they travelled against gravity in the expanding universe. They are still arriving from that distant time after having travelled a distance of 13.7 billion light years. They make an amazing image, called the surface of last scattering. This is the earliest baby photo of the universe, when it was just 380,000 years old.

The all sky image maps the tiny wrinkles which turned into the galaxies of today, and encodes the quantity of matter, dark matter and radiation in the universe. It is impossible for telescopes to see beyond the surface of the last scattering. However Einstein's gravitational waves will be able to look past this barrier to the earliest moments of the birth of the universe.

The image mapped by the Wilkinson Microwave Anisotropy Probe shows tiny variations in the temperature of space. Like a map of the spherical Earth this map of the whole sky reveals the structure of the early universe.

Wrinkles and
blemishes

Surface of last scattering
perfection
with wrinkles and blemishes
wrinkles
coding structure
defining the future
blemishes
imprint of creation

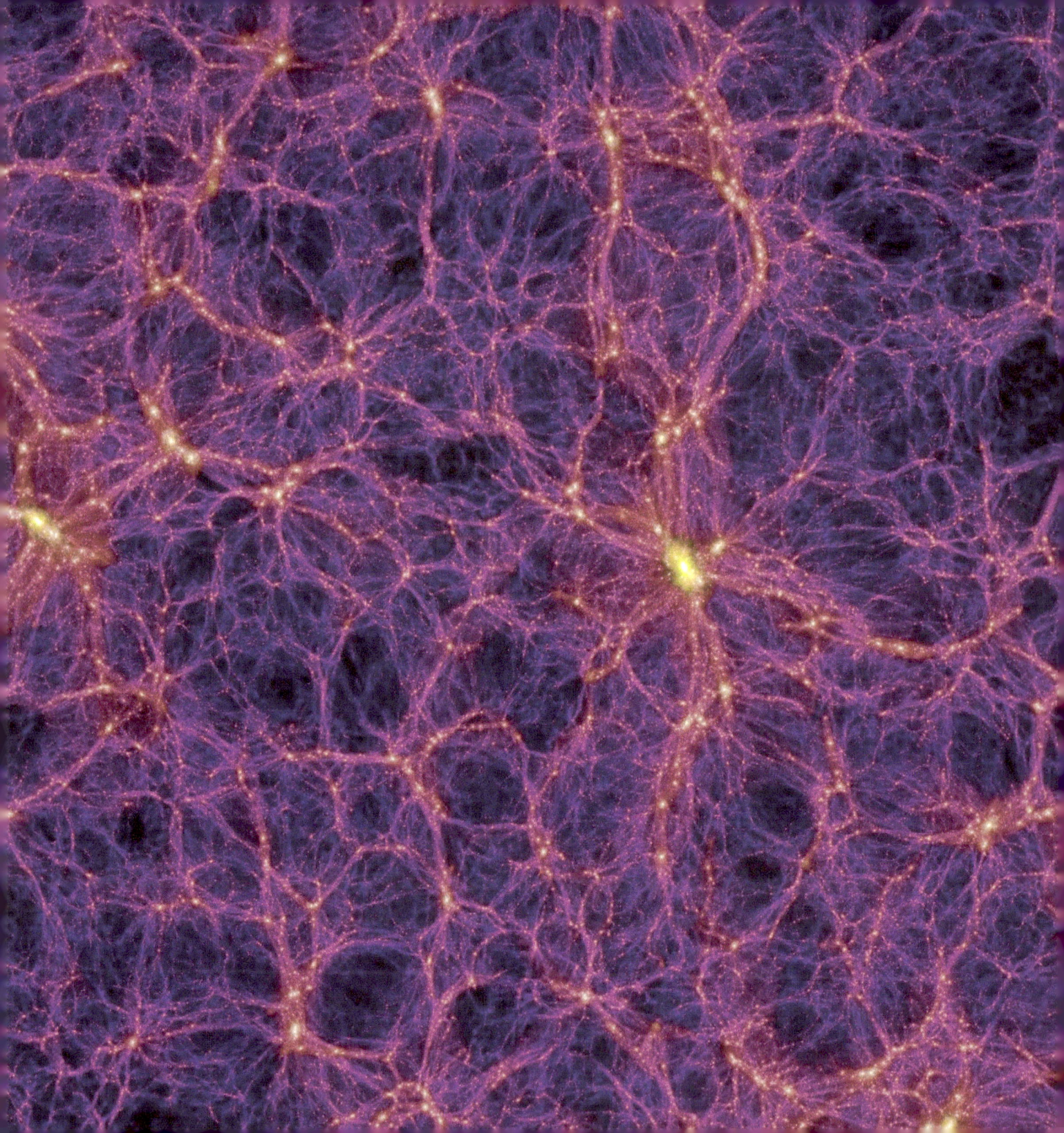

Structure Formation

Time: about 1-200 million years after the big bang, 13.7-13.5 billion years ago.

The earliest image of the universe – the surface of last scattering – shows that the universe was not quite uniform. Some patches were heavier, others were lighter. The heavier patches had stronger gravity. They attracted matter from the lighter patches. Gradually the heavier clumps got heavier, falling in on themselves, creating dense clumps of matter.

Most of this matter was dark matter, the unknown material that seems to surround all the galaxies we observe today. The gravity of the dark matter dragged with it the hydrogen and helium that was formed in the big bang. Through gravity's relentless attraction a network of filaments, clumps and empty voids emerged. The modern universe was starting to take shape.

The image shows a computer simulation of the early universe in which gravity has created hot dense clumps, filaments and voids.

Filaments and voids

Gravity
weakest of forces
sucking sucking
clumps growing
filaments emerging
voids expanding
faster and faster
webs of matter
galaxies for life

The Dark Ages

Time: about 1-100 million years after the big bang,
13.7 billion years ago.

The now transparent universe continued to expand and cool
until the sky looked black. It was as hot as an oven and filled
with gas and dark matter. Gravity was relentlessly compressing
the denser clumps of gas. There were still no stars. The
universe was dark and featureless for many millions of years.
How long the dark ages lasted is still unknown.

The first stars shone in space like a flaring match in the dark.

Darkness and heat

Darkness and heat
unobserved
seeds of structure
for future observers

The End of the Dark Ages

Time: about 200 million years after the big bang, 13.5 billion years ago.

As clumps of matter grew under the influence of gravity, denser patches within the clumps exerted their own gravity, allowing smaller patches to collapse in on themselves. As the collapses continued, the increasing gravity caused the collapses to accelerate. More collapse, more gravity, faster collapse. The universe was starting to form into a network of filaments and clumps of matter separated by empty voids. Suddenly the first stars, huge and heavy, began to form in the densest clumps of gas. They burnt rapidly.

The lights of the universe turned back on. Scientists think that the first black holes formed in this era. We know very little about the time that marks the end of dark ages. The faintest objects in the image could be amongst the earliest galaxies to have formed at the end of the dark ages.

This is an image of the Hubble Ultra Deep Field, a portrait of the most distant galaxies in the visible universe. The image features 10,000 galaxies in a tiny patch of sky.

Collapse

Gravity
drives collapse
drives gravity
drives collapse

First Supernovas

Time: some millions of years after the big bang, about 13 billion years ago.

The first massive stars probably formed in pre-galactic objects about as heavy as a million suns. The first big stars burnt fast and ended their lives in giant supernova explosions after only a few million years. As they burnt they created carbon, nitrogen, oxygen and other heavier elements. When they exploded, they enriched space with new atoms and filled it with magnetic fields. Shock waves from supernova explosions trigger the formation of new stars. Some supernova explosions leave behind black holes.

Gigantic black holes billions of times as heavy as the sun have been detected in the cores of galaxies. These are thought to have formed from the merger of black holes from the earliest supernova explosions at the end of the dark ages.

This supernova, named SN 1987A, spotted in 1987 by astronomers in Chile was the first supernova to be visible to the naked eye for almost 400 years. The supernova occurred in the Large Magellanic Cloud, a nearby galaxy visible in the southern sky. The expanding debris has now collided with a shell of gas around the star, creating this beautiful "string of pearls".

Supernova

Fat stars
living fast
dying young

fireworks

fertilising space
with
atoms for life

Growing Galaxies

Time: about 1 billion years after the big bang, 12.7 billion years ago and continuing today.

The first galaxies may be beyond the reach of our telescopes, Most of our knowledge is based on theory and computer simulation. The first galaxies were tiny compared with today's big galaxies. From computer simulations, galaxies seemed to form first at the densest clumps where filaments of matter intersected. Bigger galaxies with stronger gravity grew by attracting smaller galaxies. Many beautiful structures in the sky are a result of such galactic collisions. Galaxy cannibalism continues today.

The image shows a pair of spiral galaxies in the early stages of collision. One we see edge on, the other we see face on so that its spiral structure is visible. They are about 450 million light years from Earth. Both galaxies contain huge numbers of very massive stars which are forming very rapidly. The stars are being formed from the large amount of gas within the galaxies. Heated by the stars, the gas emits large amounts of infrared radiation. The galaxies are named UGC 9618.

Hunger

Cannibal galaxies
gobbling
engulfing
merging
growing

Generations of Stars

Stars burn hydrogen and helium and turn these simplest and lightest of atoms into atoms like carbon, oxygen, silicon and iron. When a star explodes in a supernova explosion it spreads all the types of atoms, from hydrogen to uranium into interstellar space. This way the space between the stars gets filled with dust and gas enriched with all the atoms needed to make planets and life.

Eventually the pressure of starlight acts on the gas and the dust, compressing it, and triggering collapse into new stars. Often a nearby supernova explosion can drive the debris to the point of collapse. When these second generation stars form, they almost always leave around them a ring of debris which gradually forms into planets.

A small, faint nebula IRAS 05437+2502 illuminated by stars. The dust clouds were created from previously exploded stars. In clouds like this new stars are forming.

Stars

Herded by gravity
excited by pressure
radiant by nature
suns of the universe

Giant Spiral and Elliptical Galaxies Evolve

Time: about 7 billion years after the big bang, 6.7 billion years ago.

From countless mergers of small irregularly-shaped galaxies, vast spiral and elliptical galaxies were created. Billions to trillions of stars are immersed in thin hot gas. They all seem to be buried in even bigger clouds of invisible, mysterious dark matter. At the heart of every large galaxy is a massive black hole. When there is enough gas around, the black hole can suck the material into discs of swirling matter. This in turn creates a hot magnetic funnel from which immense jets of plasma can spew out into the vast spaces between the galaxies. When the gas is used up the black hole lies dormant and nearly invisible except for its gravitational effect on nearby stars.

NGC 3079 is a barred spiral galaxy about 50 million light-years away, and located in the constellation Ursa Major. A prominent feature of this galaxy is the bubble forming in the very centre.

Painted by gravity

Soft elliptical clouds
feathery spirals
hiding
their violent past
hiding
their massive black holes
feeding
sucking
spewing
plasma fountains

The Accelerating Universe

Time: about 6-7 billion years after the big bang,
6-7 billion years ago.

Today it seems that a new force called dark energy is
blowing the universe apart. Instead of coasting and slowing
its expansion, measurements show that the expansion of
the universe is accelerating. Dark energy is a sort of cosmic
repulsion. It seems to acts on space itself, causing space to
expand.

When the universe was young and dense, the gravity of all
the matter overwhelmed the dark energy, but once it was big
enough the gravitational forces were reduced (because all
the galaxies were so far apart) and dark energy was strong
enough to overwhelm the gravity.

Dark energy might be a remnant of the vastly stronger force
that blew up the universe in the big bang. As dark energy
blows the universe apart at an ever increasing rate, the
galaxies become further apart, and galaxy collisions will
become rare. In the far future the universe will be cold and
empty with galaxies far, far apart. Is it clearing the slate for
another big bang?

In Vader

Emperor of darkness
dark energy

the dark side
overcoming

gravity
vanquished

Darth Vader
for ever

Formation of the Elements

Time: in the first few minutes after the big bang and continuing today.

Hydrogen, helium and lithium were created in the first three minutes of the universe. The medium weight elements like carbon, nitrogen and oxygen first began to form from the nuclear burning of hydrogen inside the first stars a few hundred million years later, and continue to be created inside stars today. The heavy elements from iron to uranium were created in supernova explosions.

About 90% of the atoms in your body were created during the big bang. The other 10% – the heavier ones – were made in stars and thrown into space by supernovas. They will all be recycled when you die. Yes, you are made of stardust.

This infrared image shows a region around the Tarantula nebula in the Large Magellanic Cloud. Infrared light reveals vast rippling clouds of dust that have been spewed into space by supernova explosions. Eventually some of this dust will become planets around new stars. From it, life may evolve.

I am stardust

Time begins
light atoms form
time continues
stars evolve
heavier atoms form
time goes on
supernovas explode
heavy atoms form
time continues
from the stardust
we evolved

The Birth of the Solar System

Time: about 9 billion years after the big bang, 4.7 billion years ago.

The solar system was probably formed from the debris of supernova explosions inside a star cluster when the Milky Way was already about 7 billion years old. Triggered by an explosion of a nearby supernova, a cloud of gas and dust formed and began to contract under its own gravity. As the cloud collapsed inwards its rotation increased. After around 100,000 years of collapse, the sun was born in the centre of a rotating disc of dust and gas.

The heat of the sun blew the gas outwards, leaving a disc of rocky debris where Mars, Earth, Venus and Mercury would form. The gravity of the bigger lumps of debris attracted others, creating lumps called planetesimals. Larger planetesimals attracted smaller planetesimals. Slowly the bigger planetesimals grew. Rings of debris like those of Saturn were slowly swept away. A few bigger lumps dominated, growing in a rain of planetesimals, eventually becoming the planets we see today.

The Pleiades, the Seven Sisters or Subaru is a constellation recognised by cultures all over the world. It is a young cluster of stars dominated by very hot and massive stars that illuminate the gas and dust between them. It is one of the nearest star clusters to Earth and is the cluster most obvious to the naked eye in the night sky. It was in a cluster like this that our solar system probably evolved.

Birth imagined

Star cluster
now gone
bright stars exploding
sevensisters/subaru/pleiades
do you resemble the birth home
of our friend
the sun?

The Oldest Rocks

Time: about 9.1 billion years after the big bang, 4.6 billion years ago.

Radioactive elements like uranium slowly decay into other elements. The time scales for the decay of different elements have been carefully measured. By measuring the quantities of radioactive atoms in rocks and meteorites, scientists estimate that the solar system is 4.6 billion years old.

The oldest rocks on Earth, just over 4 billion years old, are zircon crystals found in the Jack Hills of Western Australia. Rocks this old are rare, as the Earth is constantly being resurfaced by volcanic activity. To estimate the age of the solar system scientists study the slowly decaying radioactive atoms in meteorites.

The only rocks older than these are meteorites which were formed during the early condensation of the solar nebula. The oldest meteorites are found to have an age of 4.6 billion years.

The dark layers in the satellite image of the Jack Hills in Western Australia. Is where the zircon crystals have been found. The insert is of a zircon crystal from Jack Hills that has undergone ion microprobe analyses to establish it age.

Radioactive secrets

Older than old
slower than slow
atoms decay
continents drift
atoms betray
secret history
uncovered

Formation of the Earth

Time: about 9.1 billion years after the big bang, 4.6 billion years ago.

The Earth, like the other planets, formed by the gravitational accumulation of smaller planetesimals. The young Earth was mainly liquid due to the intense heat caused by planetesimal bombardment, combined with heat from the radioactivity of some of the atoms. Denser material like iron and nickel sank to the centre of the Earth forming the core, while less dense material formed the mantle and crust.

An artist's impression of the early solar system showing the formation of the planets. The inner circles are the denser rocky planets of Mercury, Venus, Earth and Mars.

Hard rain

Bombarded
and
bombarded
raining destruction
planet building

Formation of the Moon

Time: about 9.2 billion years after the big bang, 4.5 billion years ago.

About a hundred million years after the Earth was formed another planetary body about the size of Mars collided with the Earth. The enormous impact blew a vast mass of debris into orbit around the Earth. This then aggregated to form the moon.

The formation of the moon is an example of the chaotic processes that make planetary systems intrinsically unstable. Over enormous lengths of time planets can migrate towards their central star, or be expelled from the system altogether.

Planetary systems are dynamic and ever changing. In our solar system planetary orbits cannot be predicted beyond about 200 million years. In the more distant future Mercury may collide with Venus.

Artist's impression of the collision that formed the moon.

Selene

Born in violence
mother of all impacts

scarred
cratered

an image of
tranquillity

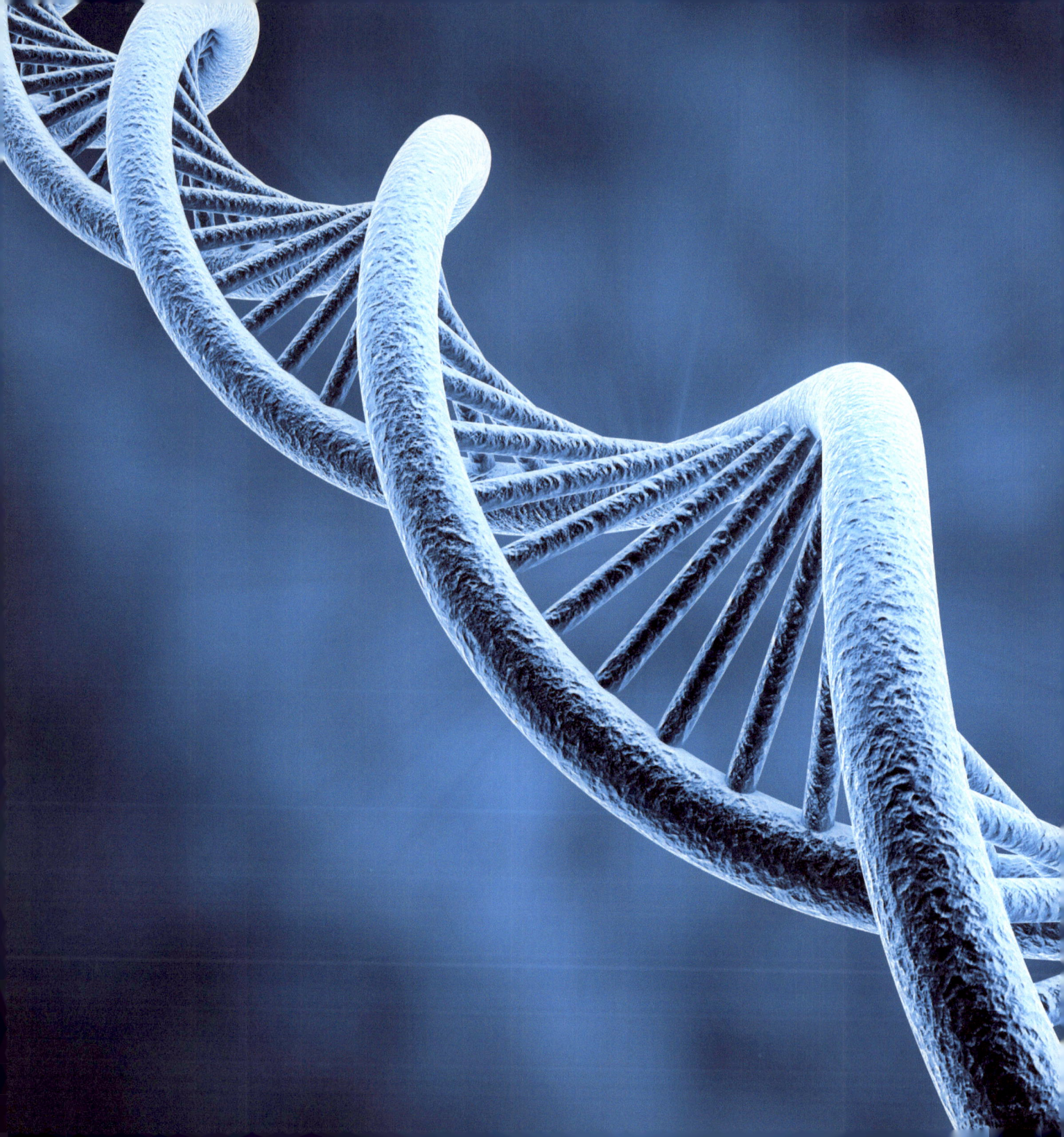

Oceans and Life

Time: about 9.6 billion years after the big bang, 4.1 billion years ago.

Planet building was still unfinished. Meteorites and comets from the icy outer reaches of the solar nebula continued to bombard the Earth. They carried water, and other molecules such as methane, ammonia and hydrogen sulphide as well as amino acids – the building blocks of life. The bombardment provided the water for oceans and puddles, as well as the compounds essential for life.

The first oceans are thought to have formed between 100 and 500 million years after the formation of the Earth. So far we do not know how the step to the first self-replicating molecules occurred. There are many theories but little evidence. Yet in only half a billion years the first common ancestor of all life had emerged, enabling the evolution of life on Earth to begin.

Deoxyribonucleic acid better known as DNA is the fundamental molecule of life. It has a structure resembling a spiraled ladder. The rungs of the ladder are made from amino acid bases, of which there are four types called adenine, thymine, cytosine and guanine. The main legs of the ladder are made from alternating phosphates and sugars. Replication happens when the molecule unzips into two half strands which are then duplicated. It is the sequence of bases along the molecule that contains the genetic codes of life.

Life begins

Hostile
hot
bombarded
cooling
solidifying
cooling
water
oceans
lightning
chemistry
DNA

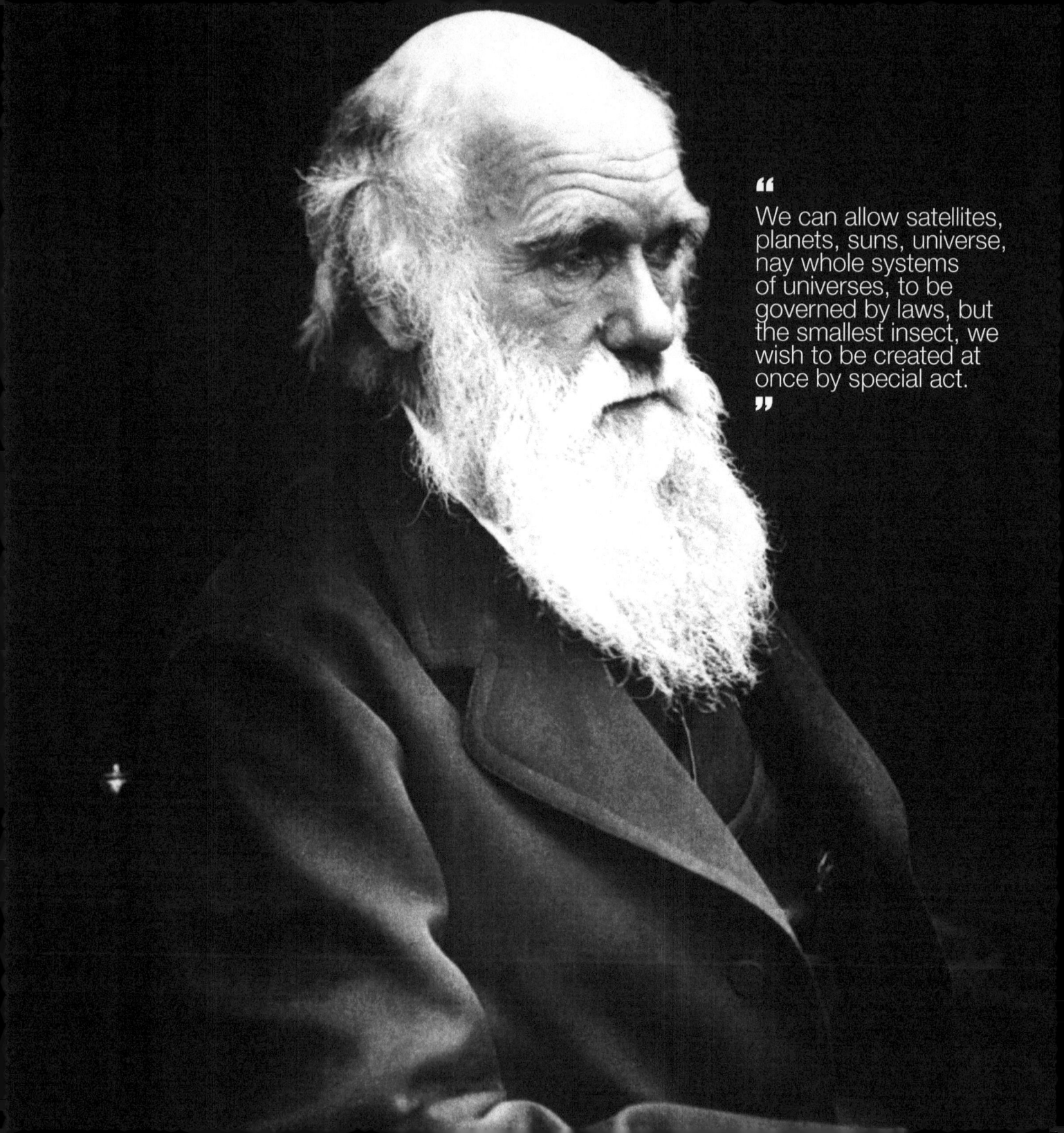

> "
> We can allow satellites, planets, suns, universe, nay whole systems of universes, to be governed by laws, but the smallest insect, we wish to be created at once by special act.
> "

Evolution and Sex

Time: 12.5 billion years after the big bang, 1.2 billion years ago.

Charles Darwin formulated a compelling scientific argument for the theory of evolution by means of natural selection. Natural selection is based on four facts. Firstly more offspring are produced than can possibly survive. Secondly traits vary among individuals, sometimes this is a result of a genetic mutation. Thirdly those with different traits have different rates of survival and reproduction. Finally trait differences are inheritable. Consequently there is a tendency for the better adapted offspring to survive and reproduce. This preserves traits that are successful. Natural selection is the only known cause of adaptation, but not the only known cause of evolution.

The evolution of sexual reproduction accompanied the transition from cells without a nucleus to the more complex eukaryote cells that have a nucleus and other structures. This occurred about 1.2 billion years ago.

Sexual reproduction shares the DNA between two parents, thereby allowing a greater diversity of genes and gene combinations for the next generation. This is very advantageous at times of rapid environmental change because it increases the chance that some of the offspring will have the necessary genes be able to cope with the change. There are still many arguments amongst biologists about the reasons for sex but undoubtedly sexual reproduction is a central process behind the evolution of life on Earth.

Charles Darwin 12 February 1809–19 April 1882.

Evolution

Chemical replication
mistake
chemical replication
life
chemical replication
mistake
more complex life
environmental pressure
survivors selected
Charles Darwin
knew

49

Death

Time: about 10 billion years after the big bang, 3.7 billion years ago.

Life is characterised by reproduction and exponential growth, but any form of exponential growth is unsustainable and leads to catastrophic overpopulation and depletion of resources, whether the population consists of humans or the first organisms on planet Earth. Thus sustainable and evolving life is impossible without death, the process that recycles nutrients and makes space for newly forming life forms. Hence we can be certain that death quickly followed the origin of life. Thus aging and death are an essential part of life.

A life span is best optimised to enable genes to survive. A few organisms can survive for more than 100 years but most have a far shorter life span. Lifetime is determined genetically. Causes of death include DNA copying errors, the switching off of error correction and mutations. The genetic causes of aging and death are not fully understood.

Eternity denied

DNA decaying
replication declining
encoded secrets fading
body aging
mutant errors increasing
chemical repair decreasing
life faltering
eternity denied

Ore Bodies Forming

Time: about 10 billion years after the big bang, 3.7 billion years ago.

The universe is characterised by the creation of structure and order. On the largest scale gravity segregates matter into galaxies and stars. On the smaller scale temperature differences and energy flows create order out of chaos. Thus complex molecules form in the dust around stars, and minerals concentrate and accumulate on planets.

Minerals concentrate and accumulate in many different ways. Life forms represent the most complex of ordered structures. They use the energy of the sun to create order and enable self replication. In many cases life forms played a major role in the formation of ore bodies. One of the most notable is calcium carbonate or chalk made from the shells of marine organisms.

Intense volcanism deposited vast amounts of metallic minerals that would become some of the great mineral deposits mined today. Early micro-organisms played a major role in their concentration as did many other different processes. At this time the seas became red with dissolved free iron that had spewed out of volcanoes. Ore bodies of gold, lead/zinc, molybdenum and copper were deposited at this time.

Sakurajima volcano located close to Kagoshima in southern Japan erupting in 2009.

Ore inspired

Pressure and heat
metals rising

volcanoes
bloody the seas

chemistry working
lifeforms helping

crystals forming
minerals depositing

veins of riches
stored away

First Life Forms

Time: about 10.3 billion years after the big bang, 3.4 billion years ago.

Around 3.4 billion years ago the first single cell life-forms (prokaryotes called cyanobacteria) had evolved from the self replicating molecules. Cyanobacteria have no nucleus and are anaerobic. They need no oxygen, instead gaining energy from sulphur based molecules. At this time there was no free oxygen in the atmosphere.

The earliest fossils of stromatolites are found in the Pilbara region of Western Australia. They were probably built by cyanobacteria that neither needed nor produced oxygen.

Stromatolites are still alive today. These are from Shark Bay in Western Australia.

First sentinels

Stromatolites
single celled
anaerobic
self replicating
builders
powered by sulphur
our first
sentinels

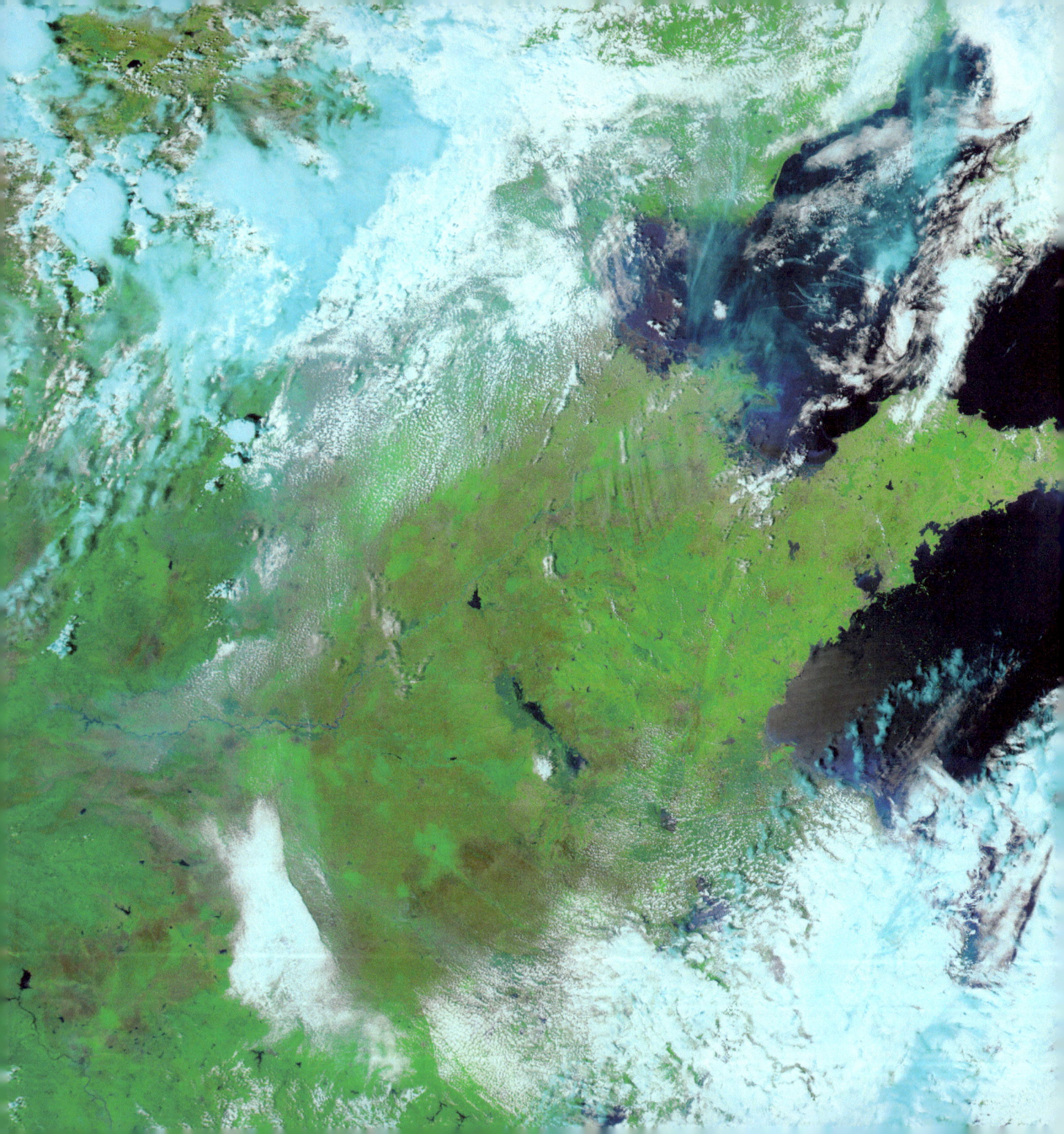

Life Discovers Photosynthesis

Time: about 11 billion years after the big bang, 2.7 billion years ago.

About 2.7 billion years ago cyanobacteria in the oceans developed the ability to produce molecular oxygen in large quantities by a process called photosynthesis. This involved capturing photons of light and using this energy to create food molecules such as sugars from carbon dioxide and water. The waste product was free oxygen that forever changed the planet.

A large cyanobacteria bloom off the coast of Ireland on May 22, 2010. The image was captured by the Moderate Resolution Imaging Spectroradiometer (MODIS) on NASA's Terra satellite.

Oxygen

Primitive
teeming
cyan

minuscule
ocean dwellers

releasers
of
Earth's
oxygen

Iron Ore Deposited

Time: about 11.1 billion years after the big bang, 2.6 billion years ago.

Around the world banded iron formations are found. These are the relics of the oxidation of the free iron in the seas that occurred when vast colonies of cyanobacteria produced the Earth's free oxygen. The oxidised iron sank to the ocean floor, changing the red seas to transparent and creating the deposits found today. These banded iron beds are estimated to have locked up 20 times the amount of oxygen present in our atmosphere today. They are the source of the rich Pilbara deposits in Western Australia.

Banded iron formation in the Pilbara of Western Australia.

Anaemia

Oceans
blood red
iron rich

bacteria
blooming

oxygen
releasing

iron ore
depositing

oceans
transparent
iron poor

Blue Sky

Time: first began about 10 billion years after the big bang, still blue today.

When photons of light pass molecules in the air, the oscillating electric field of the photon interacts with the electrons around the atoms. This causes weak scattering called Rayleigh scattering. The scattering is almost 10 times stronger for the shorter wavelengths of light – violet and blue – compared with red. When the sun shines through the atmosphere, mostly blue light is scattered in all directions. With some of its blue removed, the sun appears yellower than it really is, and the sky looks blue because wherever we look, the bluer colours are being scattered towards us.

Near sunrise and sunset, the sunlight we see comes in nearly at a tangent to the Earth's surface. The light path through the atmosphere is so long that much of the blue and even green light is scattered away, leaving the sun looking red, and the clouds illuminated in red.

In 1984 astronaut Bruce McCandless, seen here floating over the sky blue earth, made the first ever untethered spacewalk.

Scattering blue

Black night receding
sunlight arriving
atmosphere scattering
red dawn breaking
Earth turning
angles changing
blue day dawning

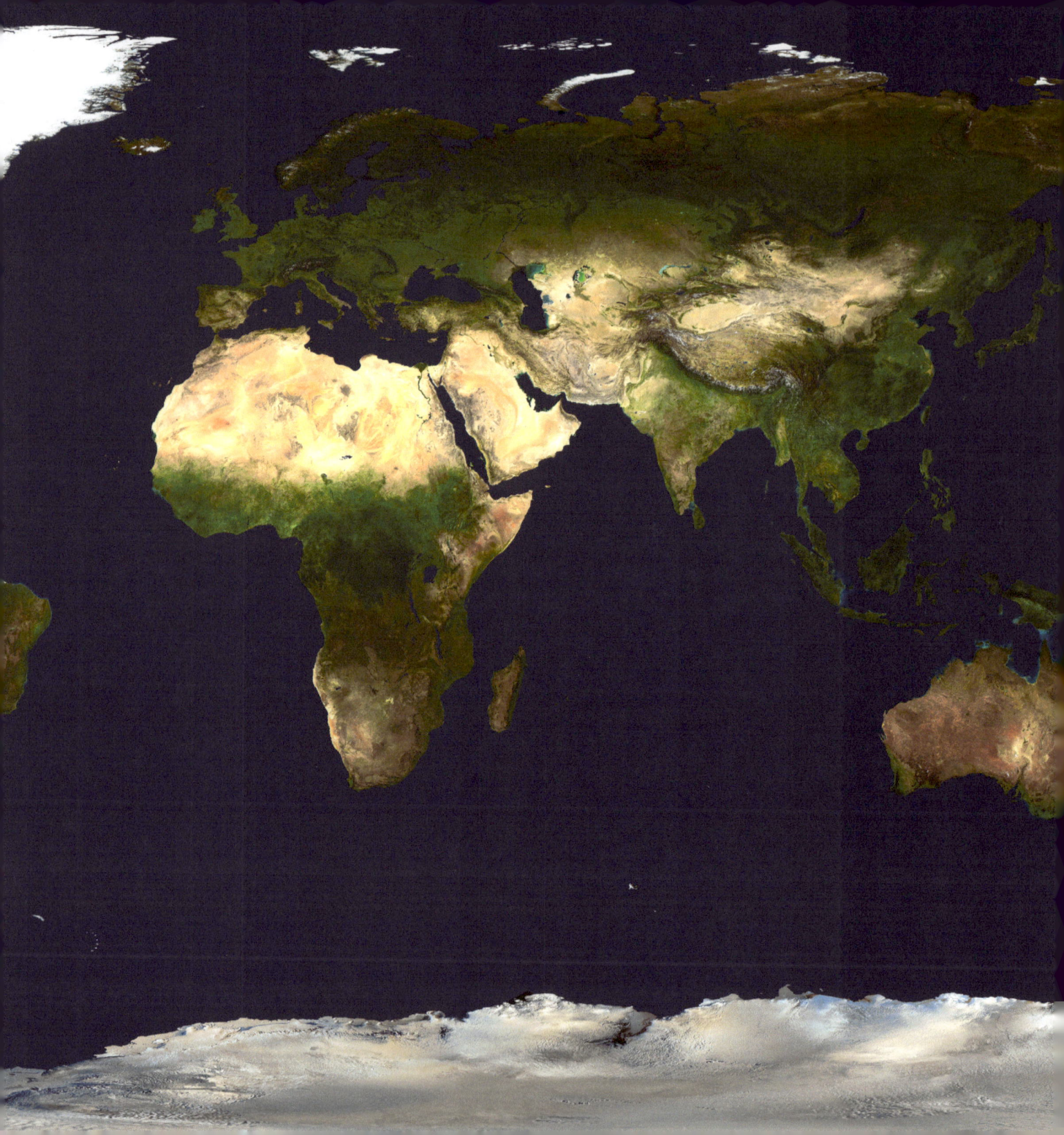

Plate Tectonics

Time: first began about 10 billion years after the big bang, continuing today.

The Earth has a thin crust of solid rock on a ball of molten iron and magma. Huge convection currents of upwelling magma rise and fall like those you can see in a bowl of miso soup. As the currents turn over they sweep the continental plates along with them at a speed of a few centimetres per year. This motion of continental plates on the Earth's surface is called plate tectonics.

Australia's motion has taken it to both the North and South Pole more than once over geological history. It is currently moving north. The edges of the continental plates, where they move under and over each other, are sites of intense geological activity, such as earthquakes, volcanoes, and mountain building.

Composite satellite image showing the position of the Earth's exposed land masses today.

Plate tectonics

Ancient surfaces
barely moving
colliding
never stopping
buckling
mountain building
never stopping
travelling
stretching
breaking
never stopping
joining
never stopping

First Record of Multicellular Life

Time: 13.1 billion years after the big bang, 600 million years ago.

The Ediacara Hills in the Flinders Ranges in South Australia contain the earliest confirmed fossils of multicellular life, including the earliest animals. They appeared towards the end of the Proterozoic Eon that ended 542 million years ago. They are known as the Ediacaran life forms. They were soft-bodied and had developed various radial symmetries.

The Phanerozoic Eon succeeded the Proterozoic Eon. The name means "revealed life" and covers the age of multicellular animal life on Earth from 542 million years ago to the present. During this time organisms left a fossil record, and built up complex and diverse ecosystems.

Dickinsonia costata, an iconic Ediacaran organism, displays the characteristic quilted appearance of Ediacaran enigmata.

Multicellular life

Two cells
link
more cells
link
sharing space
sharing living
chemical cooperation
multicellular life
begins

The First Fish Evolve

Time: about 500 million years ago.

The earliest organisms that can be classified as fish were soft-bodied jawless chordates that first appeared during the Cambrian period. Fish continued to evolve through the Palaeozoic era (542 – 448 million years ago), diversifying into a wide variety of forms.

Many fish of the Palaeozoic developed external armour that protected them from predators. The first fish with jaws appeared in the Silurian period (436-420 million years ago), after which many became very successful marine predators including the sharks.

Skull and trunk shield of the Gogo Fish (Mcnamaraspis kaprios Long, 1995). In the far north of Western Australia, near of Fitzroy Crossing, a giant fossilised barrier reef snakes across the landcape for almost a thousand kilometres. It is called the Devonian Reef. About 375 million years ago it was teeming with life. The Gogo fish was one of its inhabitants.

Fish

Scaled for protection
tailed for mobility
finned for direction
gilled for extraction
schooled in survival
buoyed in confidence
oceans to conquer

Early Land Plants Evolve

Time: about 450 million years ago.

Probably an algal scum formed on land 1.2 billion years ago, but it was not till the Ordovician period (488 – 443 million years ago), that the first land plants appeared. These were similar to modern mosses and liverworts, confined to moist environments and limited in size due to the lack of of tissue that could conduct nutrients. They began to diversify, becoming larger and more complex in the late Silurian Period, around 420 million years ago.

By the middle of the Devonian Period around 390 million years ago, most of the features recognised in plants today were present, including a vascular structure, roots, leaves and secondary wood, and by late Devonian around 350 million years ago, seeds had evolved. Plants had by now spread over most of the Earth's land mass and had reached a level of complexity that allowed them to form forests of tall trees. The spread of plants reduced the carbon dioxide in the atmosphere. This reduced the greenhouse warming of the earth and allowed an explosion of animal life forms to evolve.

Barbula Spadicea, a modern example of an ancient moss that still reproduces using spores.

Aquatic invasion

Green and aquatic
scum of the Earth

invaders of land
determined to stand

developing roots
a tenuous hold

defences evolving
existence resolving

vascular plants

First Amphibians Evolve

Time: about 370 million years ago.

The first amphibians are thought to have evolved in the Devonian period around 370 million years ago, from lobe-finned fish that had strong fleshy tipped fins. These fins allowed them to crawl out of their fresh water habitats and move between drying pools. Their air sac evolved, enabling them to breathe on land in the same way that today's lungfish do. Over time their four fins developed into limbs: they were the first four legged animals.

Modern amphibians evolved in the Carboniferous period about 340 million years ago, having adapted to live both on land and in water. Amphibians dominated the land up until a mass extinction called the Great Dying, 252 million years ago. This event may have been due to a huge meteorite impact or massive volcanic eruptions. About 96% of all marine species were wiped out, plus 70% of land vertebrates as well as many of the insect species that included dragon flies with 60cm wing span. This, together with competition from the reptiles that evolved from them, limited their numbers and range of habits. Today's amphibians include three primary groups: frogs and toads, salamanders, and caecilians.

This green tree frog has a lineage that traces back to the first amphibians.

Who am I?

With

special skin
that cannot dry

special legs
that kick and fly

special tongue that
flicks and sticks

gelatinous eggs
that look like eyes

who am I?

First Reptiles

Time: about 320 million years ago.

Reptiles evolved in the carboniferous period about 320 million years ago from the first land vertebrates – amphibians. They had adapted to living on dry land and did not require water to breed. Their soft-shelled amniotic eggs did not require an aquatic environment, as the water necessary for development of the embryo was provided within the egg itself. This enabled the reptiles to migrate away from permanent water and to continue to evolve. These early reptiles are ancestors of all later significant groups of reptiles, birds and mammals.

Male Agama sinaita, Jordan. This species is common in deserts around the shores of the Red Sea. When in heat, the male turns a striking blue colour to attract females.

Reptiles

Legs splayed
bellies low
eyes darting
sun warming
tongue sampling
the hunt beginning

Mammal Diversity

Time: about 220 million years ago.

The first true mammals evolved around 220 million years ago in the Triassic period. They were small shrew-like creatures (morganucodontids) that lived under the dominance of the dinosaurs. They evolved into a number of lineages from the early tetrapods. One line survived and from that line all modern mammals evolved. A turning point in the evolution of mammals was the mass extinction event caused by a large meteorite impact about 65 million years ago. (see page 85). The dinosaurs became extinct (except for those that evolved into modern birds) and more than half the plant species in North America were wiped out.

Most mammal species survived the meteorite impact and they were able to run, climb, fly and swim. Mammals were well insulated with fur or layers of fat. This allowed them to move between hot and cold environments. Today mammals are dominant and man is the top predator.

Man has evolved very rapidly over the last million years, developing a brain that reasons, records, analyzes, wonders and seeks answers. A graph of human population shows a catastrophic and unsustainable exponential growth. We now dominate the biosphere, we are changing the climate and we are likely to be causing a new mass extinction event. We have the power and the intellectual capacity to determine our own future but we also have a genetic legacy that drives us to conflict, domination and over consumption. Will we be able to harness our intellects to save our planet for future generations?

The African Leopard (*Panthera pardus pardus*) is a leopard subspecies occurring across most of sub-Saharan Africa. Its numbers are dwindling and its conservation status has been declared vulnerable by the by the International Union for Conservation of Nature.

Mice and men

Cheetah
fast
sloth
slow
giraffe
tall
elephant
big
mouse
small
whale
loud
platypus
venomous
man
intelligent

Insects and Flowering Plants

Time: about 130 million years ago.

The earliest known insect existed 400 million years ago. Flowers are thought to have evolved from ferns more than 200 million years ago, although the oldest fossil of a flowering plant dates from 140 million years ago. In ferns clouds of pollen emitted from male reproductive organs enabled the chance fertilisation of female eggs.

Flowers evolved to create a symbiotic arrangement between two species so that both could benefit from the services they provided. Specifically plants evolved their sexual organs into flowers to enable them to use insects and birds to assist their reproduction. The flower provides food for the insect (or bird) which in turn transferred spore or pollen to the egg of the plant to enable fertilisation.

One can imagine the first insect-aided reproduction to have been accidental. This would have given a small reproductive advantage, and a plant with a marginally more attractive (edible) sexual organ would have better reproductive success. Thus plants with tastier flowers proliferated. Once the pollinators had colour vision, there would be benefits for the flower to advertise itself with petals.

Flowering plants became dominant 60 million years ago and It is estimated that there are between 250,000 to 400,000 species of them. They are of particular importance to mankind, who has cultivated and modified many of them, especially the tubers, grasses, nuts and fruits; to provide the staple foods that we and our stock are dependent upon today. The magnificent diversity of flower forms, scents and colours has provided pleasure to mankind from his beginning.

Carpenter bee with pollen collected from a night blooming cereus plant.

Flowers

Wondrous colours
with
ingenious shapes

stamens loaded
with
pollen to spread

welcoming flyers
with
glorious scents

fuelling the workers
with
nectar rewards

Extinction

Life has evolved through countless transformations which created millions upon millions of species, most of which are now extinct. Indeed more than 99% of all life that have ever existed on Earth are now extinct.

An extinction event is a sharp decrease in the diversity and abundance of macroscopic life. Extinctions are occurring all the time but when they occur in large numbers over a short period of geological time they are known as mass extinctions.

The causes of mass extinctions are complex and not fully understood. A few have been clearly linked to meteorite impacts. Others are likely to have been caused by a combination of events such as climate change, sea level variation, and massive volcanic activity. These events, possibly combined with an ecosystem that was already under stress, could cause rapid changes in habitat, and interruptions to the food chain. Species not able to adapt to cope with the changes were wiped out.

The mass extinction of the megafauna such as mammoths in Eurasia, and other large mammals in the Americas and Australia is thought to have been primarily due to human predation.

Extinct crinoid, Agaricocrinus Americanus Carboniferous Indiana.

Extinction

Life forms
blink in
lighthouse flash
life forms
blink out
lighthouse flash
life forms
blink in
blink out

Australia Bulldozed

Time: about 300 Million years ago.

Apart from Antarctica, Australia is the lowest, flattest and the driest continent. During the Permo-Carboniferous glacial period the continental ice shelf above parts of Australia would have been around 4km thick, scouring the surface, revealing mineral deposits and forming the flat landscape we have today.

The flat lands of Lake Ballard in Western Australia featuring one of the fifty one sculptures by Antony Gormley. The castings are made from an alloy of molybdenum, vanadium and titanium, materials that are found in Archaean rocks of Western Australia.

The ice bulldozer

Endless ice moving
massive and cold

grinding and smoothing
warming and melting

endless flatness
emerging

mineral riches
revealed

The First Birds Evolve

Time: about 200 million years ago.

The first bird-like animals that could fly included the famous Archaeopteryx. They evolved from the theropod dinosaurs during the Triassic period (250 – 200 million years ago). A number of species of dinosaurs have been found to have been feathered. The evolution of modern birds is thought to have gone through a four-winged stage. Many distinct anatomical features are shared by birds and the theropod dinosaurs, including feathers and hollowed bones. Nest building, the use of gastric stones to aid digestion and brooding activities were common behavioural similarities.

Artist's impression of the earliest birds based on the feathered dinosaur fossil Anchiornis huxleyi, discovered in north-eastern China.

Flight

Theropods
changing

hollowing bones
making it lighter

scales to feathers
arms to wings

feathers lengthen
muscles strengthen

powering up
for flight

The End of the Dinosaurs

Time: about 65 million years ago.

An asteroid or comet, 10 to 16 kilometres in diameter, struck Earth in what is now the Yucatán Peninsula of Mexico. This catastrophic impact is thought to have produced tsunamis and a super heated cloud of dust and water vapour that moved rapidly around the world, causing massive fires that contributed to the dust from the impact, blackening the sky and creating darkness.

With very low levels of sunlight reaching the surface of the earth the temperature dropped, creating a global winter and restricting plant growth. These major climate changes, together with intense volcanic activity prior to the impact, caused a mass extinction of 17% of the earth's families.

This was the end of the dinosaurs except for the feathered type that evolved into modern birds. The extinction of the dinosaurs allowed mammals to diversify and grow in size during the next Era, the Cenozoic (65 million years ago to the present).

The asteroid contained iridium. Today this rare metal can be found in a thin layer deposited by the settling dust at many sites around the world.

An artist's impression of the meteorite impact believed to be the major cause of the dinosaur's extinction.

Dinosaurs dying

Impact
shattering

shockwave
passing

darkness
descending

dinosaurs
dying

Temperature and Sea Level Variation

Time: over the last 500 million years.

The temperature of the Earth has never been constant. Over the last 500 million years it has fluctuated from cold glacial periods (ice ages) to hot non glacial periods. The cycles of temperature and sea level variations are closely linked. During cool glacial periods more water is contained as ice on land, causing a lowering of sea level.

Gradual climate change is caused by many factors including the cyclic wobble of the spin axis of the Earth, changes in the Earth's orbit, the position of the continents, variations in the power of the sun and changes in the greenhouse gas composition of the atmosphere.

Sea levels have varied by up to 250 metres above present levels to 130 metres below. At present, most fresh water is concentrated in the polar ice caps. Today it is melting and the sea levels are rising.

We depend on an extremely thin fragile layer of atmosphere to support life and separate us from space. Here we see the International Space Station above the thin layer of sky. This photograph is taken from the Space Shuttle Endeavour.

Climate change

Earth's axis processing
cyclic and slow
Earth warming and cooling
another ice flow

solar intensity
waxing and waning
Earth warming and cooling
seas rising and falling

continents forming then breaking
apart
solar absorption varies with drift
Earth warming and cooling
a piece of the puzzle of climate shift

burning the carbon
warming much faster
heading towards
a manmade disaster

Modern Humans Evolve

Time: about 200,000 years ago.

Homo erectus (Latin: "upright man"), evolved around 2 million years ago and successfully migrated out of Africa around 1.8 million years ago. Modern humans, Homo sapiens (Latin: "wise human" or "knowing human") evolved in the African savannah from the Homo erectus in a transition that was completed between 100,000 and 200,000 years ago. It is thought that Homo sapiens migrated out of Africa between 70,000 – 50,000 years ago, moving across existing land bridges and seas to populate the world. Modern human behavior developed during this period with the specialisation of tools, rituals such as burials with gifts, rock art and refined hunting techniques.

The need to understand our environment has always aided our survival. Modern human consciousness was born on the African savannah, but has continued to develop, and perhaps to accelerate, as technologies such as writing have allowed thoughts and ideas to be shared across the planet and across the time of recorded history. Technology has allowed us to realise our need and our yearning, to understand our place in the universe. Technology has also given us powers to alter our planet and powers to alter our own evolution. Modern medicine and modern communication technologies are likely to dramatically change the future of human evolution. However our knowledge and our ability to reason are balanced against survival instincts born on the savannah. Reason tells us that we live on a tiny fragile planet but instinct tells us to take and to consume as though our planet was infinite. Human evolution is reaching a critical point which could lead to a transition to sustainability or to an imminent catastrophe.

A diorama in National Museum of Indonesia, Jakarta, depicting a life size model of stone equipped hunter, a Homo erectus family living in Sangiran about 900,000 years ago.

Ancestry

Homo habilis
handy man

homo erectus
bipedal man

homo neanderthalensis
strong man

homo sapiens
intelligent man

homo futurist
space man

Human Painting

Time: about 30,000 years ago.

Humans began painting at least 30,000 years ago during the last ice age. In Australia, rock paintings known as The Bradshaws were painted around 20,000 years ago. The art typically shows small, slender, decorated figures, often with bent knees and apparently performing corroborees.

Opinion is divided as to whether or not the paintings were made by ancestors of the present Aborigines or by a different race that is now extinct. They are found in the Kimberly region of Western Australia and are known by their current aboriginal custodians as Gwion Gwion rock art.

A Bradshaw painting from the Kimberley region of Western Australia.

Gwion Gwion

We were old
with
secrets untold

we were foretold
with
traditions to uphold

we were bold
with
images to unfold

The First Farmers

Time: about 10,000 years ago.

Around 10,000 years ago in what is now known as the Middle East, land clearing and farming were becoming established in the lands that were to give rise to the first permanent villages. These villages produced surplus food, enabling the first cities to develop, and they in turn enabled civilisations to develop. Some of the greatest of these arose in the fertile lands along and between the Tigris and Euphrates rivers including the Sumer civilisation. (3500 – 2334 BC)

The Middle East was not the only birthplace of agriculture. There is evidence that agriculture was developed in Papua New Guinea at about the same time that it began in the Middle East. Agriculture was also developed independently, but somewhat later, in the Americas, leading to the development of cities and empires.

A satellite image of the mouths of the Tigris and the Euphrates, Shatt al Arab Delta, Iraq-Kuwait border today.

The first farmers

Wandering gatherer of grain
wandering hunters of game
pause on a fertile plain
corral the game
plant the grain
and
remain

Discoverers, Discoveries
and the Future of the Universe

Humans evolved under dark skies on the plains of Africa. Driven by a need to understand their environment they sought understanding. They saw stars, planets, comets, meteors, the Sun and the Moon. Trying to make sense of the unreachable realm of the sky, their observations were woven into a rich tapestry, stories that linked their existence to cosmic powers. They observed how the sun and the moon controlled seasons and tides, the rhythms of migrations and fishing and crops.

Undoubtedly the objects in the sky had powers beyond human understanding, and it did not require much extrapolation to endow them with human emotions and an ability to influence lives. Stories elaborated from their observations created a new universe of mythology. They learnt the power of careful observation and recording. Records of annual events like solstices, and special events like eclipses of the sun and the moon enabled predictions of future events and gave power to those with the knowledge.

Over time people discovered ways of improving their powers of observation. The invention of telescopes, microscopes and other instruments allowed the history and majesty of the cosmos to be slowly revealed. The advances in our understanding came from careful observation combined with intellectual struggle. In the next section of the book we pay tribute to a few of the people who helped unravel the history of the universe.

Through studying the past we can extrapolate to the future. Observations today allow us to imagine the fate of the Earth and the fate of the universe. But our understanding is far from complete. The possible futures are bleak but they are also uncertain. The choices seem to be cold and empty oblivion, or an accelerating rip in which all structures, even atoms are torn apart in a final singularity in which all distances diverge to infinity!

How it all began is also uncertain because, despite the discovery of the Higgs boson, we have a deeply unsatisfactory theory of matter that fails to explain why the universe prefers matter to antimatter, why all the forces have their particular strengths, and why the universe has three dimensions of space plus one of time. We can be certain that people will continue to strive to find answers to the big questions, but whether we will ever be able to find all the answers is the biggest question of all.

Artist's concept of the Juno spacecraft launched from Earth in 2011. It will arrive at Jupiter in 2016 to study the giant planet from an elliptical, polar orbit. Juno will repeatedly dive between the planet and its intense belts of charged particle radiation, coming only 5,000 kilometers from the cloud tops at closest approach.

"
At rest, however,
in the middle of
everything is the sun.
"

Heliocentric Solar System

Nicolaus Copernicus 1473–1543.

Aristarchus, a Greek astronomer, was the first to propose that the planets revolved around the sun. Aristotle and Ptolemy believed that the sun and the planets revolved around the Earth. The Earth-centric belief held sway for the next 2000 years until Nicolaus Copernicus revived the sun centric theory of Aristarchus. His book entitled "On the Revolutions of the Celestial Spheres" is frequently referred to as the beginning of modern astronomy.

His heliocentric model, with the sun at the centre of the universe, demonstrated that the motions of celestial bodies can be explained with the sun in the centre of the planetary system. His work stimulated further scientific investigations, becoming a landmark in the history of science that is often referred to as the Copernican Revolution. Johannes Kepler and Isaac Newton further developed our modern understanding of the solar system.

A portrait of Nicolaus Copernicus housed in The Regional Museum in Toruń.

Copernicus

The
ancients
observed
their
complex prance
Copernicus
unravelled
their
elliptical dance

"
In questions of science,
the authority of a
thousand is not worth
the humble reasoning
of a single individual.
"

The Father of
Modern Science

Galileo Galilei 1564-1642.

In Galileo's time the teachings of the ancient Greek philosophers were accepted without question. Following Aristotle, it was believed that objects fell at a speed proportional to their mass, even though rather trivial experiments could easily have been used to disprove this doctrine. Galileo used experimental observations to criticise Aristotle's theory of gravity. In Pisa he discovered the universality of free fall and did numerous experiments with pendulums and inclined planes to investigate the motion of objects in the Earth's gravity. This was the start of the scientific revolution driven by a spirit of scepticism and experimental observation.

Galileo was the first person to look into space with a telescope. He made careful astronomical observations of the moon, sun spots, the planets and their moons. His support of the heliocentric views of Copernicus brought him into conflict with the Catholic hierarchy who convicted him of heresy and put him under house arrest. In 1992 the Catholic Church formally cleared Galileo of any wrongdoing.

Galileo

Ancient doctrines
medieval myths

challenged by observations
confronted by scepticism

a stubborn
search for truth

Edwin Hubble Discovers the Expanding Universe

Edwin Hubble 1889-1953.

Edwin Hubble was the first person to measure the distance of distant galaxies, thereby proving that the universe extended far beyond the Milky Way. He found that distant galaxies were receding from us at a speed proportional to their distance, an idea that had been suggested by Georges Lemaître who had used some of Hubble's own data to make his case. Lemaître had suggested that the universe was expanding in accord with Einstein's equations. It was Hubble's data that proved this idea conclusively.

The rate of expansion of the universe is described by a number that we now call the Hubble constant. This famous number specifies the expansion speed for any given spacing of galaxies. Although Hubble's data proved the expansion, his estimate of the expansion rate was about 7 times too large. The Hubble Constant has now been refined to about 20 km per second per million light years. The value of the constant is still not precisely known because galaxies also have large random orbital motions due to their attraction to other galaxies. Still, the universe is so large that the most distant galaxies are seen to be receding at speeds well over half the speed of light.

The expansion of the universe tells us that in the past the universe was much more compact. When telescopes look to the distant universe, they are seeing a much younger universe, because the light takes billions of years to travel to us. The distant universe we see with telescopes clearly shows it to be much more compact in agreement with the theory.

The Hubble Space Telescope, named after Edwin Hubble, has extended Hubble's legacy by imaging the most distant galaxies in the universe.

Hubble

Looking beyond
our milky way

receding galaxies
moving away

red shifted light
informs by night

Einstein's convenient
oversight

expanding universe
historically smaller

a pointed beginning
infinitesimally smaller

"
The most incomprehensible
thing about the world is
that it is comprehensible.
"

A New Way of Thinking

Albert Einstein 1879-1955.

Einstein was central to the development of the two fundamental theories of modern physics, general relativity and quantum mechanics. General relativity describes space, time and gravity. It leads to a completely new understanding of the universe on large scales. His theory describes how space, time and matter are connected. Quantum mechanics asserts that light comes in *quanta* called photons, but that everything has a wave-particle duality – sometimes things act like waves and at other times they act as particles.

Quantum mechanics leads to a complete re-assessment of the nature of reality, and exposes an intrinsic weirdness beyond our intuitive understanding. In particular it tells us that everything is in some sense random and statistical, and that interactions are not local – a photon here can somehow influence a photon that is far from here. Einstein was most disconcerted by this quantum weirdness. He said "God does not play dice" and he described it as "spooky action at a distance".

To this day the weirdness of the quantum world continues to baffle physicists even though the mathematical theory works perfectly. The theory of general relativity has also been spectacularly successful. Yet there is a big problem: the two theories are mutually incompatible. We still search for a unified theory that can overcome this contradiction.

Einstein

Light speed
constant
matter and energy
equivalent
space
curved
energy
quantized
Einstein
prophet of our time

Curved Space
Warped Time

Einstein's general theory of relativity is encapsulated by the saying

matter tells space how to curve
space tells matter how to move.

Newtonian physics has it that space and time are absolute and matter has no effect on either of them.

Einstein's idea that matter curves space has been proved to exquisite precision, and every time you use a GPS Navigator you use technology that only works because the engineers have corrected for the effect of matter on time. Indeed gravity is best understood as the tendency for matter to follow the fastest route through space and time. Free fall is this fastest route and gravity is the force you must apply to prevent free fall, thereby creating tiny amounts of slowing of time.

In 1976 NASA launched an atomic clock into space to measure the warping of time by the Earth. In 2011 NASA announced the results from a spacecraft called Gravity Probe B that measured the shape of space around the Earth. Both experiments proved Einstein to be right. Time depends on your height above the Earth. The magnitude of this effect is a few microseconds in an hour at 10,000 kilometres height, and the curvature of space around the earth causes the perimeter of a circle around the Earth to differ from its Euclidean value by 28mm.

Move to the time warp

Matter
moulder of space
warper of time

innocent bystander?
no!

culprit
creator
maker

space curves
time warps
and matter moves

Freefall

If you were born on an asteroid you would think that free floating in space was the natural thing. You would have no sense of down and you would not have our concept of gravity. Because we live in a special place in the universe – a solid planet with rock to hold us up – we think of falling as something very different from floating around in a space station.

In fact both ideas are the same: falling from a tower or floating in space are just the same thing: matter following the curves of space. The main difference between floating in space and falling is that when you fall the planet gets in the way. This is the bit that hurts.

Floating free

Riding the curves of space
riding the warps of time
floating
drifting
freedom falling?
some call it falling
freed
from barriers
from the crunch of matter

Time Dilation

The warp of time is very small around planet Earth. The gradient of time is about three nanoseconds per year for every meter change in height. According to Einstein's theory, and proved in many experiments, time runs fastest if you are in free fall, such as floating around in the space station. This is described as the principle of maximal ageing.

Time runs fastest, and hence everything ages the fastest when it is in free fall. Whenever you apply forces to stop free fall then time runs slower: your journey in time is extended. For us, planet Earth conveniently holds us up, and prevents us from falling freely towards the centre of the Earth. This slows time, slowing our ageing compared with an astronaut in space.

While the slowing of time by the Earth is tiny, you would need to go to a neutron star for your aging rate to be slowed significantly. There you could get eight extra years of life compared to your friends at home. At the horizon of a black hole time comes to an end: there is no observable passage of time.

Simulated view of a black hole in front of the Large Magellanic Cloud.

Time stands still

Time sprints
for floaters in space
time slows
for walkers on planets and
swimmers on stars
time drags
for explorers of neutron stars
time stands still
frozen for no one
at the shimmering horizon
of a black hole

Trinity, the First Atomic Bomb

Time: July 16 1945.

Physicists came to understand the processes that power the stars in the first decades of the twentieth century. It was soon realised that this power could be harnessed as a source of energy and also in the form of a bomb.

The development of atomic weapons arose out of the politics of the 1930s. The formation of fascist governments and the fear that Nazi Germany was working to develop an atomic bomb lead the United States, Britain and Canada to begin a secret program called the Manhattan Project. This culminated in the testing of a nuclear weapon at what is now called the "Trinity Site". The atomic bombing of Hiroshima and Nagasaki followed just a few weeks later.

The first atomic bombs harnessed energy stored in the nuclei of heavy atoms created in supernova explosions. In the 1950s weapons scientists worked out how to use an atomic bomb to trigger a much more powerful explosion created by the fusion of hydrogen nuclei. This is the power that drives the sun. The enormous destructive power of hydrogen bombs motivated a worldwide campaign that led to international agreements to reduce arsenals that many believe were big enough to cause a mass extinction event. The partial success of this campaign gives hope that humans need not destroy themselves with their own technology.

The trinity explosion, 0.016 seconds after detonation. The fireball is about 200 metres wide.

Trinity a chain reaction

Powerful minds
conceived it

the Manhattan Project
built it

political consideration
unleashed it

a chain reaction
of destruction

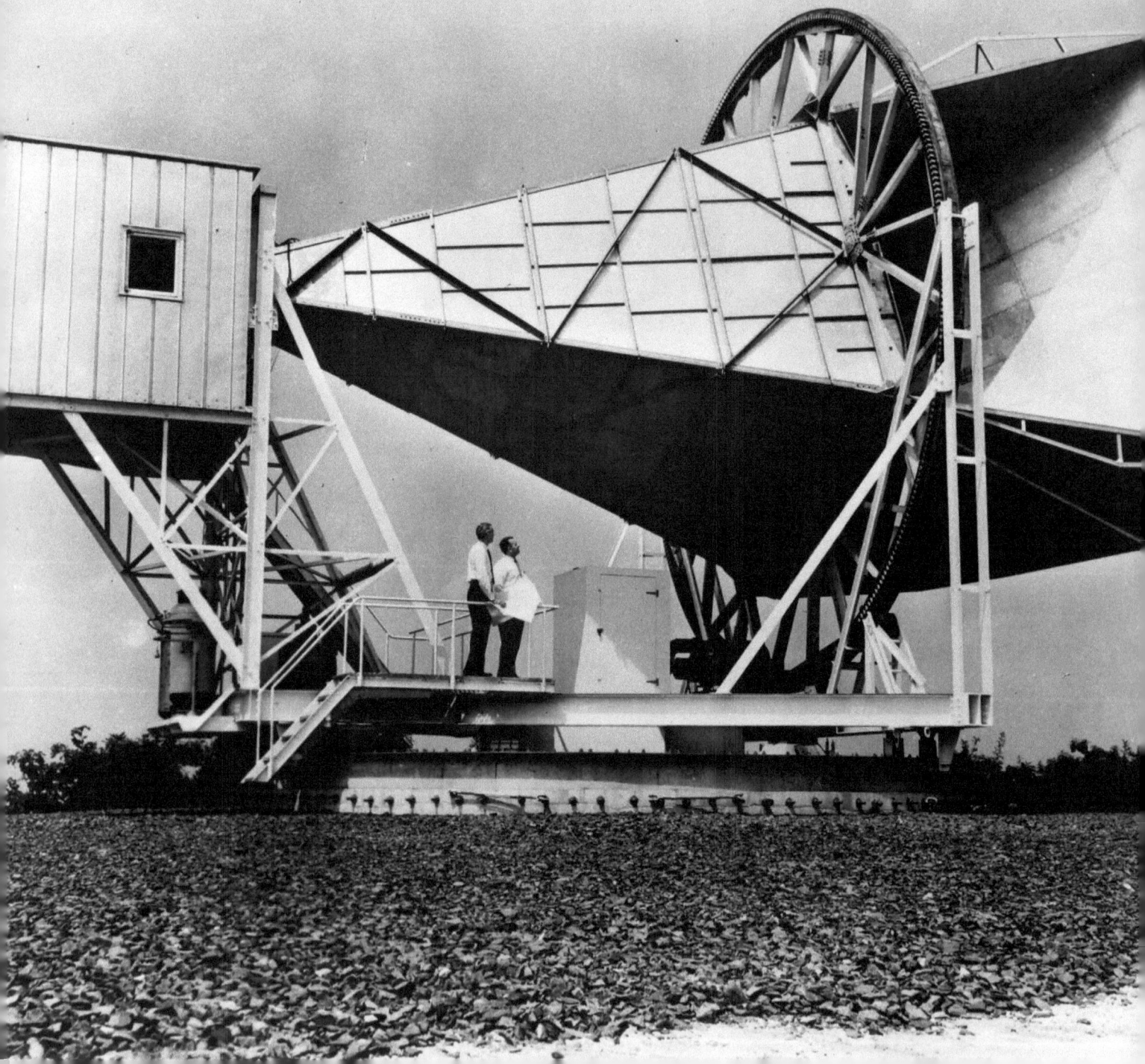

The Cosmic Microwave Background

Time: 1964.

The idea that photons from the big bang would form a background of low energy photons coming from all directions in the sky was first predicted by George Gamow in 1946. In 1964 American and Russian scientists detected this weak glow from the big bang.

It was soon realised that this glow of microwaves (described on page 19), carried information that would allow detailed study of the big bang. It represents an image of the universe as it looked when it was 1000 times smaller and 1000 times hotter than it is today.

Three spacecraft have made successively better images of the background. In 1989 NASA's Cosmic Background Explorer made the first blurry images. In 2001 its successor called the Wilkinson Microwave Anisotropy Probe created the image shown on page 19. In 2009 the European Space Agency launched a spacecraft called Planck which is making even finer resolution maps. All have confirmed the big bang theory and allowed detailed measurements of the large scale properties of the universe.

Bell Labs' Horn Antenna in Crawford Hill New Jersey USA that was used by Penzias and Wilson to detect the cosmic microwave background.

Background to understanding

Once upon a time
space
was hot and shining
today
it is colder than cold
minus 270 degrees
whispering shush in radio

Dicovery of Pulsars

Time: 1967.

In 1939 Robert Oppenheimer predicted a new kind of star, called a neutron star. His theory was based on recent discoveries in nuclear physics, the discovery of the neutron and quantum mechanics. A neutron star is essentially a ball of neutrons about 20 kilometres in diameter, but as massive as our sun. The idea was extraordinary because it said you could have a star that was really a vast atomic nucleus. The idea was forgotten by most people, and Oppenheimer went on to build the atomic bomb.

In 1967 Jocelyn Bell discovered regular radio pulses coming from the sky. First she thought they might be a beacon from a extraterrestrial civilisation. Eventually these pulsating stars or pulsars were realised to be rapidly rotating magnetised neutron stars. They emit radio waves, light and X-rays in a beam like a light house beam.

With more than 2000 pulsars now discovered we know that they are created when a massive star explodes in a supernova explosion. Some pulsars exist in binary pairs with other stars. In these systems matter falling from the large star onto the neutron star causes the neutron star to be spun-up. Some have been found to spin at more than 600 revolutions per second – faster than a dentist's drill!

Pulsar systems have been wonderful systems for testing Einstein's general theory of relativity.

The Crab Nebula in the constellation of Taurus. In the core of this nebula the Crab pulsar spins 30 times per second. It looks like a star except that it flashes on and off 30 times per second. It can be easily seen through telescopes. The nebula was formed in a supernova explosion observed by Chinese, Japanese and most probably Native Americans astronomers in the year 1054 AD. The image covers a six-light-year-wide expanding remnant of the star's supernova explosion.

Pulsars

Whirling Dervishes
of space
cosmic lighthouses
in place
neutron stars beaming
at a devilish
pace

"
That's one small step
for a man, one giant
leap for mankind.
"

Man Lands on the Moon

Time: July 1969.

Launched from the Kennedy Space Centre on July 16, Apollo11 was the fifth manned mission and the third lunar mission of NASA's Apollo program. The Commander Neil Armstrong and the Lunar Module Pilot Buzz Aldrin became the first humans to walk on the Moon on July 21. Their Lunar Module, Eagle, spent 21 hours 31 minutes on the lunar surface, while Command Module Columbia pilot Michael Collins remained in orbit. The three astronauts returned to Earth on July 24, landing in the Pacific Ocean.

Lunar Module Eagle descending to the Moon.

1969

Sea of Tranquillity
desolate
undisturbed
until
an eagle
carried by
Apollo
landed

Gravitational Waves

Einstein's theory of general relativity predicts a new sort of wave – gravity waves – which like light and radio waves carry information from the distant universe. Gravity waves are created when dense matter accelerates. Vast bursts of gravitational wave energy are predicted to be released when black holes form and when pairs of black holes coalesce.

Throughout our Milky Way galaxy close pairs of neutron stars have been discovered that appear to be losing energy and spiralling together in accordance with Einstein's prediction of gravitational wave emission. Eventually they will collapse into a black hole, making a characteristic chirping sound in gravity waves.

Gravitational waves can pass through all matter. Hence gravitational wave detectors can see into the heart of an exploding star, and to the birth of the universe itself. They offer the hope of explaining the incompatibility between the theories of gravity and quantum mechanics.

Around the world physicists have built huge observatories that aim to detect these elusive waves. They hope to detect the gravitational waves created in the chaotic turbulence of the big bang, as well as the pure vibrations of black holes in the moments after their formation. In this way gravitational waves may reveal the beginning of time in the big bang, and the end of time in the black holes that are the gravestones of dead stars.

An aerial view of the Virgo gravitational wave detector located near Pisa in Italy. The L-shaped structure, of 6km total length, consists of huge vacuum pipes that contain intense laser beams and finely suspended mirrors that aim to measure the tiny vibrations of space when gravitational waves pass through the Earth.

Gravity waves

Fabric of space
stretched over time
rippled by mass
history encoded
in
gravity waves

Self Awareness

The forces of the universe have transformed it from being hot, uniform and structureless to one where structures occur on a huge range of scales, from vast clusters of galaxies to nanometer sized molecules.

 On at least one tiny planet matter has evolved into a form we call life, in which infinitesimal inhabitants of the universe (we call them "us") have been able to observe the universe and map out its history and structure. Does the universe have an innate ability to create observers able to look at the whole? Are we privileged to be part of a lucky accident or are we part of an inevitable process that is happening all over the universe?

The image shows a 50 light-year-wide view of the tumultuous central region of the Carina Nebula where a maelstrom of star birth – and death – is taking place.

Know thyself

Atomic fuzz
matter flows
patterns evolve
chaos defeated
information rules
pattern creates pattern
matter evolves
universe
did you always know
you would
know yourself

Consciousness

The driving force for life is the temperature difference between our sun and empty space. As shown by Ilya Prigogene, temperature differences are the fundamental drivers for creating order out of chaos. The heat flow from the sun to the Earth's surface, and from there into space leads to the ultimate expression of an ordered and organised system – the conscious human brain. The enormous variety of processes that ultimately lead to ordered living organisms, Prigogene calls negentropic processes.

Each of us is a chance collection of atoms driven together by a complex web of such negentropic processes. Our atoms are created in stars and in the big bang. However, we do not own our atoms. They are in a state of flux, coming and going, with a turn over time of a few months. Yet these same atoms encode our DNA and the messages in the DNA are almost immortal! Some of our genes are billions of years old. Thanks to the gigabytes of information encoded in our DNA, and repeated in every cell, and thanks to the cooperation of all those billions of cells, we are able to be conscious of ourselves and our place in the universe.

Each of us is an incredibly precious and special product of chance, and yet totally insignificant in the oblivious vastness of space.

Conciousness is epitomised by the meditative awareness of our own minds. This wooden statue of Buddha is from the Chinese song Dynasty (960-1279), at the Shanghai Museum.

Conscious universe

Chance on chance
chance collections of chance
collections
of matter
emerging
from fire behind
and
freeze before

here one moment
gone the next
awareness remaining
awe and humility
within
oblivious vastness

Forces in Nature

The standard model of particle physics recognises four fundamental forces, the strong and the weak nuclear, the electromagnetic and the gravitational force. The strong force holds atomic nuclei together. It has to be immensely strong to overcome the electric force which makes the positively charged protons in an atomic nucleus repel each other.

The weak nuclear force causes radioactive beta decay. It is associated with a special type of charge called leptonic charge which is carried by electrons and neutrinos.

The electric force is associated with both electricity and magnetism: magnetism is explained by Einstein's theory of Special Relativity as the effect of motion on an electric charge. The electromagnetic force has two charges, usually called positive and negative. Unlike charges attract and like charges repel.

The weakest force of all is the gravitational force. It is a force exerted by all forms of matter and energy, and it acts on space and time to change its curvature. The change in the shape of space causes matter to move. In the words of John Archibald Wheeler "matter tells space how to curve" and "space tells matter how to move".

Gravity is in a sense illusory, because if you are floating freely in space you cannot feel it. It is the force you need to apply to prevent matter from following the curves of space. For gravitation there is only one charge: it is always positive so that there is no such thing as gravitational repulsion.

The Large Hadron Collider at the European Centre for Nuclear Research, CERN, seeks to understand the forces of nature and the structure of matter by smashing protons together to discover their basic constituents. This photograph is a view of the 7000 tonne ATLAS detector that is placed around one of the collision points. It measures the debris produced when protons collide head-on at a speed fractionally less than the speed of light. In 2012 the Higgs particle was discovered at CERN.

The four forces

Mr Strong
binding twine
strapper of bundles
confiner
of quarrelling protons

Mr Weak
knot undoer
saboteur
atom breaker
neutrino escapee

Electroman
photon friend
two faced
life friend
always straight
magnet man

Gravity
crushing mirage
overwhelming illusion
cloaked
by the warp of time
in the fabric of space

Are We Alone?

Is there extraterrestrial life, or are we alone in the universe? While many people have made claims, there is no scientific proof of the existence of alien life – no repeatable observations and no material evidence. Amino acids, the building blocks of life, have been found in meteorites. This means that the chemical processes that lead towards life began in space. If life began in space, then meteorites and comets might carry life across the galaxy, bringing life to every suitable planet soon after it forms.

The search for extraterrestrial intelligence tries to detect radio signals or laser flashes from alien civilisations on distant planets. Some hope that there may be aliens out there broadcasting enough television, radar beams or mobile phone signals that we could pick them up. The amount of energy required to be detectable by our existing telescopes is huge: much greater than the amount we currently broadcast.

Other enthusiasts hope for deliberate beacons in the form of narrow beams directed at our planet. Such beams would allow all the civilisations in the Milky Way to share knowledge and explore their galaxy without leaving home. A galactic club out there may be trying to encourage new members. But there is a catch! Since the discovery of radio a spherical bubble of radio emission from Earth has expanded less than 100 light years, encompassing a tiny fraction of our galaxy. Only very nearby aliens can tell that a radio communicating civilisation exists here, assuming they have suitably vast radio telescopes. If they have just detected us, their reply will not arrive for another 100 years.

This alien looking image is a southerly looking night view of the upper two-thirds of the Florida peninsula taken from the space shuttle. If aliens could detect our city lights they could guess that the Earth had technological life.

Are we alone?

Alive
inquisitive
we wonder
are
we alone
I
hope not
or
are we
being watched
I
hope so

Milky Way and Andromeda Collide

Time: about 16.7 billion years after the big bang, 3 billion years into the future.

Milky Way and Andromeda galaxies are on a collision course. In about 3 billion years, the two galaxies will collide. Then after one billion years of a very complex gravitational dance, they will merge to form an elliptical galaxy. During this period, the gas in these galaxies is likely to be compressed to trigger a star burst.

For perhaps one hundred million years stars will be formed hundreds of times faster than normal, so that the galaxy lights up with clouds of hot blue stars.

The galaxies shown here are located 300 million light-years away in the constellation Coma Berenices. The colliding galaxies have been nicknamed "The Mice" because of the long tails of stars and gas emanating from each galaxy. Otherwise known as NGC 4676, the pair will eventually merge into a single giant galaxy.

Galactic collision

Their gravity
linked them at birth
then
forced their circular dance
It will merge them
excite them
and
make them one

The Earth is Vaporised

Time: about 21.3 billion years after the big bang, 7.6 billion years into the future.

Our sun will start expanding when its hydrogen is used up and its helium begins to burn in a shell away from the core. The change in the balance of heat generated pressure and gravity will cause the sun to expand into a giant red star, like Betelgeuse in the constellation of Orion. It is predicted that the earth will be swallowed and vaporised by the swollen red sun in about 7.6 billion years. The Earth will be seared at more than 3000 degrees. All life on Earth will end.

Vanquished Earth

Sol our star
aged and dying
expanding
hydrogen spent
helium burning
red and gigantic
engulfing
consuming
death to
mother
Earth

End of the Stellar Era

Time: trillions of years in the future.

The first stars formed around 100 million years after the big bang and stars will continue to form until 100 trillion years after the big bang. Already today star formation is much rarer than it was in the early universe. After a few trillion years star formation will be very rare.

By this time most of the hydrogen and helium that fuelled the stars will be gone, converted to heavier elements. Star formation will slowly cease and the remaining stars will slowly fade, leaving dim and cooling stars. Darkness will be descending.

This is an artist's concept of the red dwarf star CHRX 73 and its companion CHRX 73 B.

The dimming

Celestial light dimming
recycling ending
black holes engulfing

endless expansion

overwhelming
our once
majestic universe

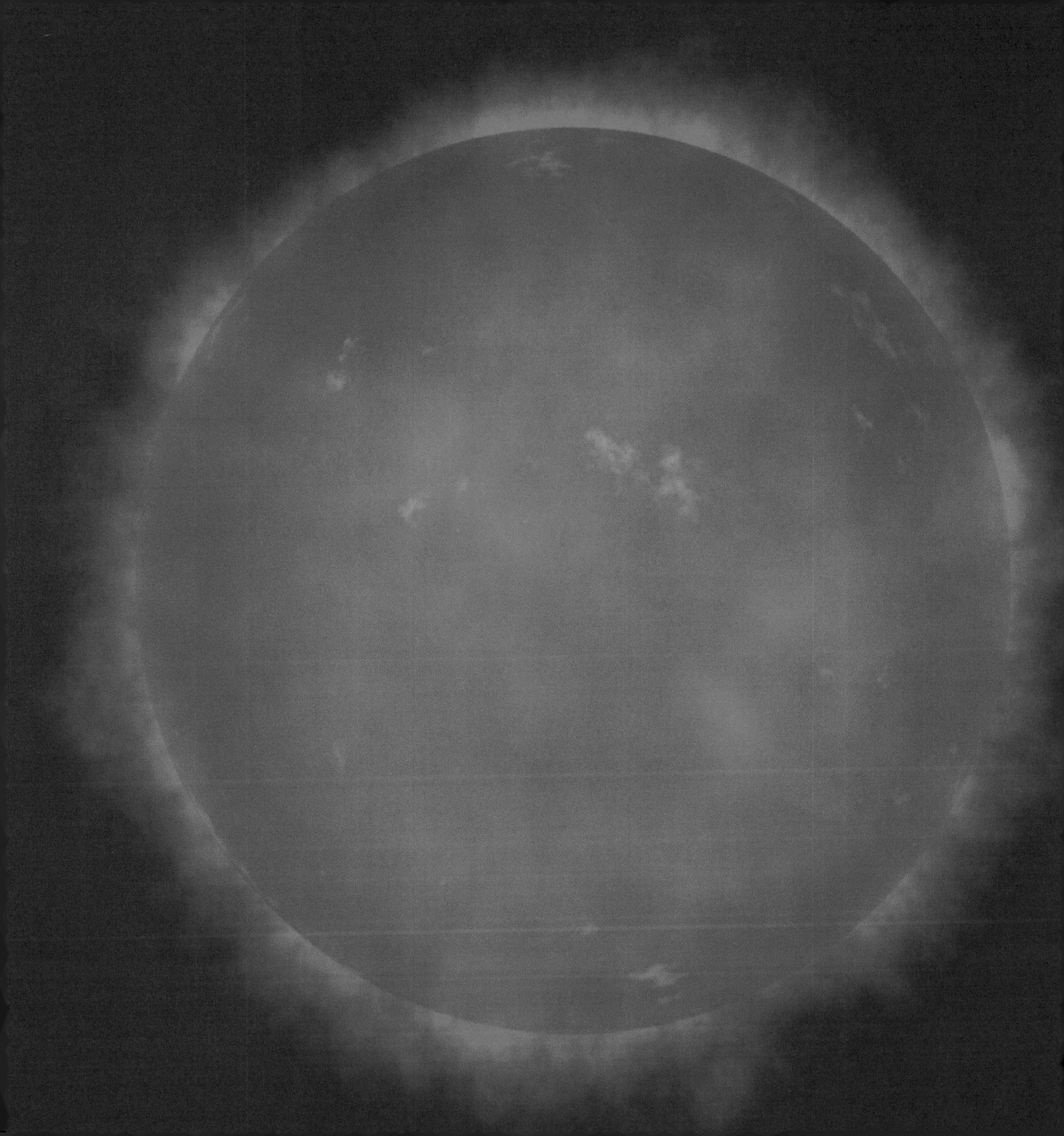

The Degenerate Era

Time: hundreds of trillions of years in the future.

The slowest burning stars are red dwarfs. They can burn for hundreds of billions of years. New stars will continue to form slowly, but by hundreds of trillions of years into the future star formation will have finished and the last stars will have died.

All that will remain will be the cold corpses of four types of stars: red dwarfs, white dwarfs, neutron stars and black holes as well as many cold dark planets.

Once the last of the red dwarfs have exhausted their fuel, all nuclear fusion in the universe will have ceased. The universe will be a very dark place.

Artist's conception of star SO25300.5+165258 which is a red dwarf about 12 light years from the sun.

The last star

Ancestors gone
I am alone
elementally heavy
energy waning
light diminishing
slowly freezing
a feeble ember
fading
into
infinite darkness
I am
the last star

The Black Hole Era

Time: about 10,000 trillion trillion trillion years (10^{40} years) in the future.

After the embers of the white dwarfs, red dwarfs and neutron stars have faded, around 10^{40} years from now, when all protons and neutrons have decayed into positrons and neutrinos, black holes will dominate the universe. (According to some theories protons are actually unstable and over enormous periods of time they can decay.)

During this era, the primary source of radiation throughout the universe comes from black holes. They cannot live forever; they slowly evaporate by an emission called Hawking Radiation. Small fluctuations in space-time cause particle-antiparticle pairs to appear close to the event horizon of a black hole. Due to the strong gravity so close to the black hole, one of the pair is captured while the other escapes into space. This feeble signal will be the only source of warmth throughout the cosmos. The universe will be colder and darker than it has ever been.

Black holes like this artist's impression will continue to grow while there is matter around, but far in the distant future they will evaporate and dissipate to nothing.

A dark riddle

Infinitely small
Infinitely dense
a
singularity
a
quantum contradiction

Tasks for the Future

Physics today is built out of two fundamental theories: the theory of quantum mechanics which describes interactions of matter at the micro level, and Einstein's General Theory of Relativity which describes the large scale properties of space, time and gravity. Unfortunately these two theories are mutually incompatible. Most physicists think that there must be a unified theory that can combine everything...a theory that explains the properties of space, time and matter at all scales from the tiniest sub-atomic particle to the universe as a whole. This theory should describe the universe from its moment of creation to the far distant future.

To understand the smallest scales physicists have developed a quantum theory for the particles that make up atomic nuclei. This theory, called The Standard Model had a great victory in 2012 when the one missing link, the Higgs Boson, (predicted 40 years earlier), was discovered. Unfortunately, the Standard Model itself is not very satisfactory. It depends on 21 seemingly arbitrary numbers to describe the various particles.

Behind all the successes of physics are big unknowns. We have no idea why the universe has the particular properties it has. If properties like the speed of light or the electromagnetic force were to be changed by just a little bit, the universe would be radically altered. For example stars could not produce the atomic elements, or they would burn up so fast that there was not time for life to evolve. The universe seems to be fine tuned for life. A theory that can answer these deep questions seems to be very far away.

The image shows simulated particle tracks in the ATLAS detector on the Large Hadron Collider (LHC) at CERN, which discovered the last missing particle of the Standard Model of Particle Physics in 2012. While completing the theory of particle physics it leaves enormous questions.

Limited understanding

Wanting
ultimate understanding
but
needing
a
quantum leap
into
relativity

The Big Rip

Dark energy seems to be a kind of pressure that is blowing the universe apart. Observations today support the idea that dark energy is a property of space itself. Every cubic meter of space has a certain amount of dark energy and as space expands the total quantity of dark energy increases. However, as space expands the matter in the expanding space is diluted. The more the matter is diluted the easier it is for dark energy to overcome gravity and expand the universe even faster. More dilution gives faster expansion gives more dilution and so on.

As this accelerating expansion takes place first the galaxies are ripped away from each other by the expanding space. Then the stars will be ripped away from the galaxies. Then, with space expanding even faster, the atoms will be ripped away from the stars and finally even the atoms will not be able to withstand the expansion of space. They too will be ripped apart.

Because our knowledge of dark matter is still imprecise we cannot be certain that this is the ultimate fate of the universe. However, this theory which is based on dark energy being a constant property of space predicts that the universe will become infinitely large and infinitely diluted within a finite time in the future. Another consequence of this idea is that the size of the observable universe shrinks as the expansion accelerates because the distance to places that are expanding faster than the speed of light becomes smaller and smaller. In the end all parts of the universe become disconnected. The big rip is a final singularity in which all distances diverge to infinity and everything is ripped apart.

A ripping yarn

Space and matter
mystery moment
expansion started
for a moment

expanding space
expanding matter
defined by matter
for a while

happy balance
equal partners
dance together
for a while

hidden dark side
overcoming
matter losing
for a while

Space expanding
darkness growing
matter tearing
all for nothing
ever more

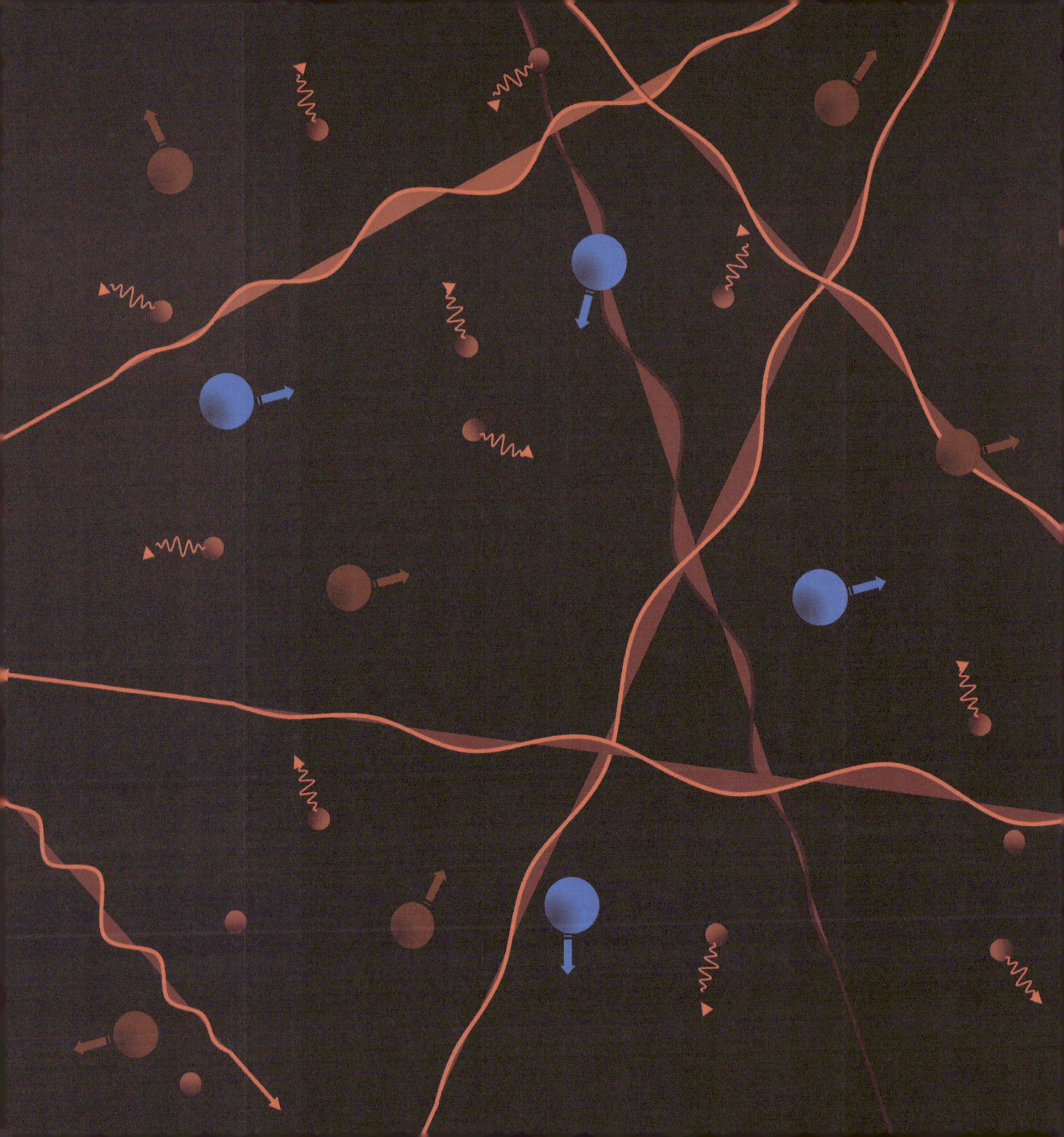

The Big Freeze

Time: About 10^{100} years into the future.

Dark energy may not be a constant quantity as discussed on page 141. If dark energy is dissipated as the universe expends, we are offered a completely different future. Instead of being ripped to shreds, the universe in this case will continue to expand steadily and for ever.

We can consider a vastness of time beyond our comprehension. In this far future the stars will have burned themselves out. If it is true, as predicted by some theories, that protons are radioactive and can decay into radiation, then we can expect that all matter will be converted to radiation. All the dead stars in the galaxies will radiate away and even the black holes will have radiated themselves away by Hawking radiation (see page 137).

We will end up with a universe that consists of nothing but a vast sea of electrons, positrons, neutrinos and radiation. As expansion continues, the radiation will cool. The universe will be vast and empty, cold and dark. For all intents and purposes the universe as we know it will have dissipated.

Artist impression of the remnants of the universe at its demise.

Endless ending

The coldest cold
the darkest dark
older than oldest
density infinitesimal
wavelength infinite
universal oblivion
dark energy's job done

An Awesome and Humbling Story

Humanity is bound together by our common yearning to understand our place in the universe. Our knowledge of the universe has vastly expanded in the past 100 years. We know much about what happened and when, and we know why many things happen: why stars explode, and why life forms evolve. We have discovered an awesome and humbling story. Yet we still do not know answers to the most basic questions. Luckily we still have a long way to go. In a universe without questions we might be bored to death!

Artist's impression of the 5km diameter central core of SKA antennas. The Square Kilometre Array (SKA) is a radio telescope which is planned to have a total collecting area of one square kilometre. With receiving stations distributed 3000 kilometres out from its core it will be the most sensitive pair of radio telescopes ever built. The construction is scheduled to start in 2016 in both Australia and South Africa. The project that aims to answer fundamental questions about the origin and evolution of the universe.

Why?

Why, why, did you begin?
How, how did you expand?
Symmetric perfection of nothing:
You broke it into forces
You dictated the
dominance of matter.
Why did you have to give space
its dark repulsion?
Why must we float in
dark matter unseen?
Are observers like us
everywhere?
necessary?
to bless creation with humble awe?
Or are we an accident?
alone?

About David Blair

David Blair's passion for science was switched on by the first Sputniks. He built radios and rockets and robots, and spent a summer with NASA during the Apollo missions.

After completing a PhD thesis called Superflow which was about the amazing properties of superfluid helium, he searched for a real challenge and chose to join the early efforts to detect Einstein's predicted spectrum of gravitational waves. Today his passion is still the detection of Einstein's waves. The field has grown from a few physicists in a few university laboratories, to a discipline with more than 1000 physicists working in billion dollar projects, building detectors of unimaginable perfection. The waves remain elusive, yet the beauty of Einstein's theory tells us that they must be there.

David is director of the Australian International Gravitational Research Centre. His passion for communicating science was fulfilled with the concept of the Gravity Discovery Centre , that today sits beside the Gingin gravitational wave research facilities. With his friend and mentor, the late John Delaeter, he developed the concept of an education centre that combined art with science, and traditional knowledge. The centre brought artists, scientists and religious and cultural leaders together to create a place that celebrated the unity of the human yearning to understand our place in the universe, and the rich diversity of our cultural explanations.

David's goal was to ensure that everyone whatever their background, should have the opportunity experience the awesomeness of the universe. For this reason the Gravity Discovery Centre supports education at all levels. From university courses to pre-school excursions, its artworks and exhibitions are designed to allow people to experience the sense of awe we all feel when confronted by the vast grandeur of the universe. The poems in this book were written to try to encapsulate key ideas, to catch the feel of key moments in the vast process of creation, to put the ideas outside the realm of scientific jargon, so that

David Blair Verse

even the most unscientific person could get a feel for the story of creation as revealed by modern science.

David has written science plays for children, using comedy and time travel to teach the evolution of fundamental ideas that ended with Einstein's revolutionary discoveries. He co-authored a popular book called Ripples on a Cosmic Sea, about the quest to discover Einstein's waves, (which was translated into 5 languages). He created many sculptural exhibits to illustrate the ideas of Einstein's universe at the Gravity Discovery Centre. He is an author of about 500 research papers and has edited several books and conference proceedings, the latest of which is Advanced Gravitational Wave Detectors published in 2012.

About Geoff Cody

Geoff Cody's first picture with his first camera captured the circular trails left by the southern stars above the South Pole using a long exposure. His childhood was spent exploring and collecting from the rivers, ocean, and bush around his home in Perth.

His fascination with science continued at school where an enthusiastic science teacher "Doc McKenna" had built a radio receiver that enabled the students to tune-in to radio signals from the first Earth satellites. The "Doc," as he was affectionately known, inspired Geoff to study physics, chemistry, geology and philosophy at university and to start a deeply rewarding forty year career as a science teacher.

He has developed many innovative science programs that have inspired students to be inquisitive innovative and to develop a deep love of the sciences.

Geoff's fascination with science inspired him to lead the creation of a 60 metre long installation called *Timeline of the Universe* for the Cosmology Gallery at the Gravity Discovery Centre in Western Australia.

The timeline included poems that tried to encapsulate the ideas and times in the history of time. Following numerous requests from members of the public who were inspired by their poetic efforts, Geoff and David were persuaded to convert this work into a book.

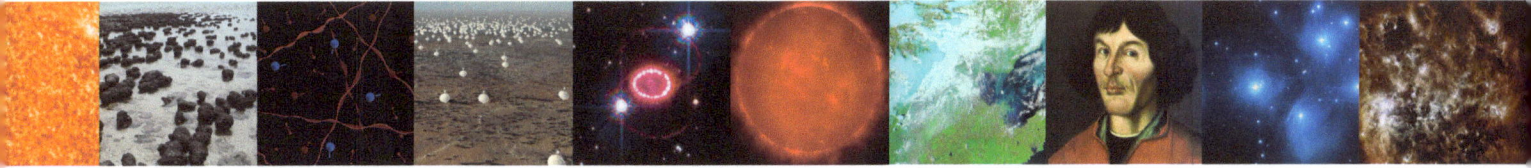

Geoff Cody Verse

Image Credits